Electrical Power PE Practice Exam
Table of Contents

Table of Contents

Click on each line below when viewed from your computer to go directly to that page:

1. Introduction……………………………………………………………..3
2. How to Use this Practice Exam………………………………….4
3. Question Distribution……………………………………………..6
4. Useful Articles and Cheat Sheets……………………………..7
5. Answer Sheet…………………………………………………………..9
6. Morning Session Questions……………………………………13
7. Afternoon Session Questions…………………………………37
8. Morning Session Solutions……………………………………..60
9. Afternoon Session Solutions…………………………………124
10. Answer Key……………………………………………………………187
11. Question Order by Subject……………………………………189
12. List of Qualitative Questions…………………………………193

Electrical Power PE Practice Exam
1. Introduction

Introduction

My name is **Zach Stone, PE**. I have been helping hundreds of engineers just like you pass the PE exam since 2014. This practice exam is part of our online program at www.electricalpereview.com which contains over 400 videos of worked out problems, 450 practice problems with full solutions, a complete online webinar live class program, and reference material to print out for every subject on the exam.

This practice exam is intended for educational purposes only and is not to be used for the actual practice of engineering in any way shape or form. **Electrical PE Review, INC** is not affiliated with **The National Council of Examiners for Engineering and Surveying (NCEES®)**, the non-profit organization that develops, administers, and scores the Principles and Practice (PE) examination. The NCEES® website is located at www.ncees.org.

Each question in this practice exam is designed to help prepare you for the NCEES® PE exam, and is based on the most recent NCEES® exam specifications.

Have a question or found an error? I read and respond to every email I receive. Please feel free to contact me directly:

zach@electricalpereview.com

How to Use This Practice Exam

Depending on how far along you are in your preparation for the PE exam, there are two different ways you can use this practice exam.

The first, if you are just starting to study, is to use this practice exam as just a large volume of practice problems to study and learn from.

The second, is to treat this like the real PE exam by completing it in an eight hour sitting with a one hour break in between the morning and afternoon session. This is the preferred way to use this practice exam since it will help get you accustomed to solving problems you are unfamiliar with, help you become more familiar with using your references, and prepare you for the pace required to work problems for the extended duration of the eight hour PE exam.

To take this practice exam like the real PE exam, consider the following:

- Go somewhere quiet that is distraction free.

- Have all of your references in front of you. Not only will it help you get more familiar with what is in each reference, it will help develop your skill at quickly using your references to help you answer questions you are not familiar with.

- Set a four hour timer for each session and do not stop to check the solutions or to see if you answered correctly.

- Instead of answering each question in the order that it appears, complete each session in multiple passes by answering the easiest questions first, the medium difficulty questions second, and the hardest questions last. Taking the exam in multiple passes is one of the most successful test taking strategies for the PE exam.

- On your first pass, If you find yourself reaching for a reference book to help you answer a question then you should most likely skip the question and save it for the second or third pass unless you know exactly what it is you are looking for, and where it is located like a specific formula you are already familiar with.

- The last 15 to 30 minutes on your four hour timer should be spent trying to answer the last handful of the more difficult questions in the session with the help of your reference books.

Electrical Power PE Practice Exam
2. How to Use This Practice Exam

- Grade your exam only after you have completed the entire exam over an 8 hour time period. On a separate piece of paper, make a list of every question you got incorrect, including questions you had to make an educated guess on, even if you got it correct. Include the date that you took the sample exam.

- Over the course of the following few weeks, work through the solutions of the questions that you got wrong, and the questions you guessed on. Seek out more questions in each of those subjects (the quiz questions in our online review course is a great place to start).

- Several weeks after you've completed working through the solutions, try attempting the same set of questions you made a list of that you got wrong (or that you had to guess on) during the first complete attempt of this practice exam. When you are able to answer each of these questions correctly, cross it off the list. Consider doing this every few weeks to narrow down the list to learn from repetition and to keep challenging yourself with the problems you find difficult so that you retain the material.

For more information on advanced test taking strategy for the PE exam that should also be applied to this practice exam, please watch the free three-part video series "**How to Pass the PE Exam**" at:

https://courses.electricalpereview.com/courses/free-how-to-pass-the-pe-exam-the-very-first-time.

If you are not already enrolled in the paid version of our online program for the Electrical PE Exam, then **I strongly recommend signing up for the Free Trial of our online review course**.

All you need to sign up for the **Free Trial** is a valid email address. In exchange, you'll gain access to additional practice problems with worked out solutions, demonstration videos of worked out problems, and printable reference material that you can take with you to the PE exam. It will also give you a good idea of what to expect with our paid program and help you decide if purchasing it is the right decision for you.

To sign up for the Free Trial, click on the following link: Electrical PE Review - Enroll in the Free Trial. Alternatively, you can enroll in the Free Trial by going to our website and clicking on **"Start Now for Free."**

Also worth reading is the article "*What Do Successful Engineers That Pass the Electrical Pe Exam Have in Common?*"
https://www.electricalpereview.com/successful-engineers-pass-electrical-pe-exam-common/

Electrical Power PE Practice Exam
3. Question Distribution

Question Distribution

The question distribution for the **Electrical PE Review - Electrical Engineering PE Practice Exam** matches the most up to date NCEES exam specifications::

Ch. 1   Measurement and Instrumentation......………...4 Questions

Ch. 2   Applications......…………......………......……..8 Questions

Ch. 3   Codes and Standards......………...……………..12 Questions

Ch. 4   Analysis......………......……….....………...9 Questions

Ch. 5   Devices and Power Electronic Circuits………......7 Questions

Ch. 6   Induction and Synchronous Machines.................8 Questions

Ch. 7   Electric Power Devices......…………......…………...8 Questions

Ch. 8   Power System Analysis......………....………..……..11 Questions

Ch. 9   Protection......………......………………...............13 Questions

There are 80 Questions Total.

The 80 questions in this practice exam are in randomized order to mimic the PE exam.

There is also an equal distribution of correct answer choices:

    25% of the questions have a correct answer of A
    25% of the questions have a correct answer of B
    25% of the questions have a correct answer of C
    25% of the questions have a correct answer of D

This means that if you have to make a random guess at an answer, you will have the same probability of guessing correctly, just like the actual PE exam.

Useful Articles and Cheat Sheets

The following is a list of some of our most popular and most asked for free articles located at http://www.electricalpereview.com/free-articles/.

If you are reading this PDF on your computer, then you should be able to click each of the following links below to open the article in your web browser. If you are reading a printed copy of this sample exam, then it might be easier to navigate to the article from our main website at www.electricalpereview.com by clicking on **"Free Articles."**

1. **Recommended references for the PE exam updated each exam semester:**
   https://www.electricalpereview.com/recommended-references-resources-electrical-power-pe-exam/

2. **Leading and lagging power factor cheat sheet, including why the power angle always equals the impedance angle:**
   http://www.electricalpereview.com/leading-lagging-cheat-sheet/

3. **How to avoid the most common power formula mistakes that will cost you points on the PE exam:**
   http://www.electricalpereview.com/biggest-mistake-commonly-made-three-phase-power-formulas/

4. **A detailed worked out per unit example designed to fix 99% of any issues you are currently having with the per unit system:**
   https://www.electricalpereview.com/per-unit-example-tips-tricks-watch-electrical-pe-exam/

5. **A comprehensive look at the changes to the NCEES® electrical power PE exam specifications that took effect starting with the 2018 exams:**
   https://www.electricalpereview.com/changes-ncees-exam-specifications/

6. **An in depth look at all of the different and confusing impedance terms and where they come from (resistance, inductive reactance, capacitive reactance, conductance, admittance, acceptance, susceptance, etc):**
   https://www.electricalpereview.com/impedance-resistance-reactance-inductance-capacitance-admittance-conductance-susceptance-whats-difference/

7. **An explanation of where the square root of three (√3) multiplier comes from that is used in most three-phase power calculations:**
   https://www.electricalpereview.com/square-root-three-3-electrical/

8. **The open delta transformer connection complete with circuit layouts, phasor diagrams, and an explanation on maximum power for an open delta connection:** https://www.electricalpereview.com/open-delta-transformer-connection/

9. **4 Wire High Leg Delta Transformer Connection:**
   https://www.electricalpereview.com/4-wire-high-leg-delta-transformer-connection/

10. **How to know when to use three-phase or per-phase quantities when calculating base impedance when using the Z = V²/S equation:**
    https://www.electricalpereview.com/base-impedance-single-three-phase-values-equal/

11. **Symmetrical components single line to ground fault, an in-depth worked out example with complete details:**
    https://www.electricalpereview.com/symmetrical-components-single-line-ground-fault-electrical-pe-exam/

12. **What Do Successful Engineers That Pass the Electrical Pe Exam Have in Common?**
    https://www.electricalpereview.com/successful-engineers-pass-electrical-pe-exam-common/

To see all of the free articles, cheat sheets, and in depth explanations that we currently have available, please visit: https://electricalpereview.com/free-articles/ or click on **"Free Articles"** from the top menu bar on the main homepage located at: www.electricalpereview.com.

Answer Sheet and Scoring

Electrical Power PE Practice Exam
5. Answer Sheet and Scoring

How to Use The Answer Sheet

To get more exposure to the actual testing environment of the PE exam, print out a new copy of the *Answer Sheet* on the following page every time you attempt this practice exam. Bubble your answer for each question on the *Answer Sheet* and use the space provided underneath each question in the *Practice Exam* to work out the problem.

During the exam, only the answers you bubble are graded. You do not get credit for partially correct answers, or for showing your work. You are also **not** allowed to write anywhere else besides the exam booklet that contains the questions and your answer sheet where you record your answers.

Prior to sitting for the PE exam for the first time, most are not familiar with only having limited space to work out each problem while having to record answers in a separate answer sheet.

Imagine realizing halfway through the exam that you accidentally skipped a row in the answer sheet and that all of the answers you selected by bubbling the correct letter were recorded on the wrong line for the wrong question number. Surprisingly, this is not uncommon. It helps to get familiar with using a separate answer sheet ahead of time.

Record your answers to the practice exam by filling in the bubble for the appropriate letter choice. For example, if the answer to #23 is B, use your pencil to fill in the letter B bubble:

If you guess on a question, or are not entirely sure that your answer is correct, lightly shade the bubble instead so it is easier to erase if you decide to change your answer later:

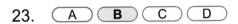

This will also make it easier to quickly identify all of the questions you should go back to review by glancing at your answer sheet if you complete the exam and still have extra time left. It also makes it easier to quickly darken in all of the lightly shaded bubbles to complete the exam if you are about to run out of time in order to ensure that all of your answers are counted.

Electrical Power PE Practice Exam
5. Answer Sheet and Scoring

Name:_____ Date:_____

Attempt #:____   # of Correct Answers:_____   Percent Score out of 80:_____

1. A B C D
2. A B C D
3. A B C D
4. A B C D
5. A B C D
6. A B C D
7. A B C D
8. A B C D
9. A B C D
10. A B C D
11. A B C D
12. A B C D
13. A B C D
14. A B C D
15. A B C D
16. A B C D
17. A B C D
18. A B C D
19. A B C D
20. A B C D
21. A B C D
22. A B C D
23. A B C D
24. A B C D
25. A B C D
26. A B C D
27. A B C D
28. A B C D
29. A B C D
30. A B C D
31. A B C D
32. A B C D
33. A B C D
34. A B C D
35. A B C D
36. A B C D
37. A B C D
38. A B C D
39. A B C D
40. A B C D

41. A B C D
42. A B C D
43. A B C D
44. A B C D
45. A B C D
46. A B C D
47. A B C D
48. A B C D
49. A B C D
50. A B C D
51. A B C D
52. A B C D
53. A B C D
54. A B C D
55. A B C D
56. A B C D
57. A B C D
58. A B C D
59. A B C D
60. A B C D
61. A B C D
62. A B C D
63. A B C D
64. A B C D
65. A B C D
66. A B C D
67. A B C D
68. A B C D
69. A B C D
70. A B C D
71. A B C D
72. A B C D
73. A B C D
74. A B C D
75. A B C D
76. A B C D
77. A B C D
78. A B C D
79. A B C D
80. A B C D

Electrical Power PE Practice Exam
5. Answer Sheet and Scoring

How to Score Your Results

If you don't pass the PE exam, NCEES® will send you a diagnostic report similar to the *Scoring Evaluation Table* shown below that demonstrates how well you performed in each subject.

After you grade your *Answer Sheet* from the previous page, use the *Question Order by Subject Answer Key* located in the back of this practice exam to fill in the **# of Questions You Answered Correctly per Subject** column in the table below.

Once you've done this for each of the nine major subjects, fill in the **Percent of Questions Answered Correctly per Subject** column by dividing the number of correctly answered questions per subject by the total number of questions per subject, multiplying by 100 to convert the decimal to a percentage.

Use this information to help determine where you should be spending your time studying in order to help improve your overall score.

Scoring Evaluation Table:

Subject	# of Questions You Answered Correctly per Subject	Total # of Questions in the Practice Exam per Subject	Percent of Questions Answered Correctly per Subject
Ch. 1 - Measurement and Instrumentation		4	0%
Ch. 2 - Applications		8	62%
Ch. 3 - Codes and Standards		12	50%
Ch. 4 - Analysis		9	22%
Ch. 5 - Devices and Power Electronic Circuits		7	14%
Ch. 6 - Induction and Synchronous Machines		8	70%
Ch. 7 - Electric Power Devices		8	14%
Ch. 8 - Power System Analysis		11	33%
Ch. 9 - Protection		13	25%

Morning Session
Questions 1 - 40

1. The phasor diagram of an unbalanced three-phase four wire system is shown below. Calculate the line current to the nearest ampere on the C phase if $I_b^{(1)}$ = 9.00A<-132°, $I_b^{(2)}$ = 3.75A<-135°, and $I_b^{(0)}$ = 4.66A<157°.

   (A) 4

   (B) 10

   (C) 17

   (D) 21

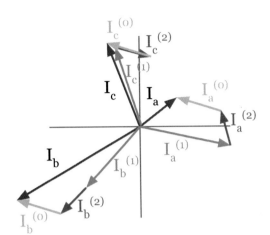

2. Select the information below that is not required to be on a motor nameplate according to the 2017 National Electrical Code®.

   (A) Rated full load motor speed.

   (B) Continuous or time rating.

   (C) Rated heater voltage if applicable.

   (D) None of the above.

3. Calculate the average DC voltage output to the nearest volt of a single-phase, 120V RMS 60 Hz full wave rectifier circuit made up of four insulated gate bipolar transistors if the gate of each pair of transistors are sent a pulse width signal that begins at 0.926 ms and ends at 7.41 ms for each period of the rectified voltage that the pair is associated with.

   (A)   87

   (B)   102

   (C)   114

   (D)   130

4. Select the minimum size copper 90 degree Celsius rated conductor run in conduit that is permitted to provide power to a three-phase, 60 Hz, 208 Volt, 60 HP, continuous duty, NEMA D induction motor according to the 2017 National Electrical Code®.

   (A)   2/0 AWG

   (B)   3/0 AWG

   (C)   4/0 AWG

   (D)   250 kcmil

5. Select the most accurate description below of how arc fault circuit interrupters operate for the application of residential protection.

   (A) Arc fault circuit interrupters protect residential circuits by de-energizing the branch circuit when the difference between the current on the hot and neutral conductor is greater than approximately zero amps.

   (B) Arc fault circuit interrupters protect residential circuits by de-energizing the branch circuit anytime a circuit overload occurs that could potentially lead to a fire.

   (C) Arc fault circuit interrupters protect residential circuits by de-energizing the branch circuit when any amount of arcing is detected.

   (D) Arc fault circuit interrupters protect residential circuits by de-energizing the branch circuit when a dangerous level of arcing is detected.

6. Select the answer below that best describes the change in trip characteristics of an adjustable circuit breaker after the time delay setting has been changed to be more inverse compared to the previous setting. Assume no changes have been made to the pickup setting.

   (A) The circuit breaker will take more time to operate for the same level of fault current.

   (B) The circuit breaker will take less time to operate for the same level of fault current.

   (C) The circuit breaker will operate at a lower level of fault current compared to the same time duration as before.

   (D) None of the above.

7. A square metal junction box is needed for the following: a single yoke three-way light switch. with three 12 AWG conductors, and a 12 AWG uninsulated ground wire terminated on the switch held in place by a cable clamp inside the box, one 12 AWG and 14 AWG conductor spliced together inside the box, and one 12 AWG conductor that passes through the box. Select the minimum box size that may be used according to the 2017 National Electrical Code® if all conductors originate outside the box.

(A) 4 × 2 ½ inch

(B) 4 × 2 ⅛ inch

(C) 4 ¹¹⁄₁₆ × 1 ¼ inch

(D) 4 ¹¹⁄₁₆ × 1 ½ inch

8. A three-phase delta connected power source delivers power to a 10+j8.2Ω three-phase delta connected load through a 0.47+jΩ line impedance. Calculate the magnitude of the delta phase voltage of the power source to the nearest volt if the current circulating in the A phase of the delta source is 55A<-17°. The entire system is balanced and positive (ABC) sequence.

(A) 240

(B) 470

(C) 507

(D) 879

9. Calculate the efficiency of a three-phase, 60 Hz, 460 V, 250 HP induction motor that draws 275 A at a lagging 0.87 power factor from a 480 volt bus.

   (A) 91%

   (B) 94%

   (C) 96%

   (D) 98%

10. Select the statement below that best describes the change that is most likely to occur to the mechanical angular displacement of a synchronous generator if the internal stator voltage begins to lead the terminal voltage by a factor of 3 while the generator is synced to an infinite bus. Assume a reference of zero degrees for the generator terminal voltage.

    (A) The change in angular displacement will cause the generator to over excite and deliver reactive power in VARs to the infinite bus.

    (B) The poles in the rotor will lead the poles in the stator field by an increase of a factor of 3 times the previous mechanical angle.

    (C) The displacement angle that the rotor was leading the stator field will now decrease by a factor of 2.

    (D) The mechanical angular displacement would not be affected.

11. Select the most appropriate response out of the choices below that accurately describes the relationship between transformer inrush current and differential protection.

   (A) Transformer inrush current is a sinusoidal component typically present only during transformer startup that is not seen by the differential protection relay due to being 180 degrees out of phase.

   (B) Transformer inrush current does not contain a DC component and can potentially cause a differential protection relay to operate due to CT saturation.

   (C) Transformer inrush current is linear in both the primary and secondary winding and does not result in mismatch differential current.

   (D) Transformer inrush current occurs in the primary winding only and can potentially cause a differential protection relay to operate due to mismatch.

12. Determine the positive sequence voltage component during a three-phase fault on a 230 kV system that has the following impedance characteristics: $Z^{(1)} = Z^{(2)} = 10$ pu and $Z^{(0)} = 2Z^{(1)}$

   (A) 132kV < 0°

   (B) 29.7kV < 72°

   (C) 28.6V < -32°

   (D) The positive sequence voltage is equal to zero.

13. Select the most accurate statement below that describes the effects of negative sequence harmonics when present on an electrical system.

   (A) Negative sequence harmonics sum at the neutral point of three-phase wye connections and can lead to the neutral conductor overheating.

   (B) Negative sequence harmonics circulate between phases and can lead to overheating conductors due to their additive nature with the fundamental frequency.

   (C) Negative sequence harmonics oppose the rotating magnetic field in motors and can cause the motor windings to overheat as the motor draws more current.

   (D) Unlike triplen harmonics, negative sequence harmonics are not harmful to the electrical system since they do not contribute to the total harmonic distortion of the system.

14. Determine how much current is seen to the nearest ampere by the ANSI #50 relay for the three-phase system represented by the single line diagram shown below during a short circuit equal to 3 times the transformer rated current if the CT's are delta connected. Assume the entire system is both balanced and positive ABC sequence.

   (A) 5

   (B) 6

   (C) 8

   (D) 10

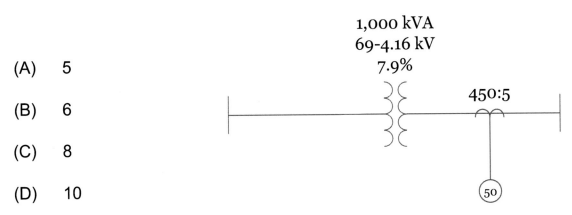

15. Calculate the total impedance in ohms of a single-phase, 1 MVA, 4,160-480V transformer referred to the low voltage side if the impedance values of the transformer are 1.20Ω primary resistance, 4.80Ω primary reactance, 0.016Ω secondary resistance, and 0.064Ω secondary reactance.

   (A) 0.032 + j0.13

   (B) 1.20 + j4.80

   (C) 2.40 + j9.61

   (D) 90.2 + j361

16. Three single-phase transformers are connected wye-delta to form a three-phase transformer. Select the correct voltage rating below of the three-phase transformer if each single-phase transformer has a voltage rating of 7.97-230 kV.

   (A) 7.97-230 kV

   (B) 13.8-230 kV

   (C) 7.97-132.8 kV

   (D) 23.9kV-230 kV

17. Select the control method from the choices below that has the highest hierarchy for preventive and protective risk according to the NFPA® Standard for Electrical Safety in the Workplace.

   (A) Personal protective equipment (PPE)

   (B) Awareness

   (C) Approach boundary

   (D) Administrative controls

18. A new motor can be purchased by paying $500 a month for 5 months with no money down. To lower monthly cost, an alternative payment plan is agreed upon that will cost $1,100 down but will reduce the monthly payment to $125. Determine the uniform monthly savings of the payment plan rounded to the nearest dollar if the annual interest rate is 12%.

   (A) $25

   (B) $50

   (C) $85

   (D) $148

19. Calculate the probability of reliable electrical power for the customer bus shown in the single line diagram below by using the equipment reliability and unreliability probability values in the given table. Assume that the generators are redundant and that each has enough capacity to provide power to the customer bus on its own.

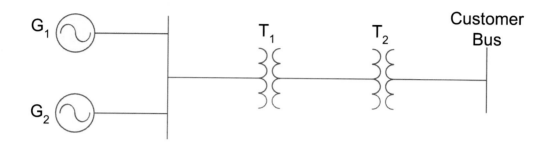

Equipment	Probability of Reliability R(t)	Probability of Unreliability F(t)
G1	0.92	0.08
G2	0.87	0.13
T1	0.96	0.04
T2	0.94	0.06

(A)   0.89

(B)   0.72

(C)   0.93

(D)   Cannot be determined.

20. Three 250 kcmil uncoated copper conductors per phase are ran in PVC conduit to provide power to a three-phase load 75 feet away from a 480 volt power source. Calculate the percent voltage drop of the system in accordance with the 2017 National Electrical Code® if the load draws 650 amps at a lagging 0.80 power factor.

(A) 0.2%

(B) 0.4%

(C) 0.6%

(D) None of the above.

21. Calculate the complex impedance (Ω) of the line in the three-phase system shown below if the system base values are equal to the ratings of the transformer.

(A) 0.11 + j0.44

(B) 0.04 + j0.15

(C) 15.4 + j90.6

(D) 30.2 + j121

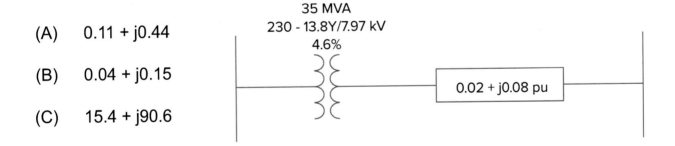

22. The results below are from an insulation time resistance test at one test voltage for a cable used in 2 kV applications. Evaluate the results based on the standard dielectric absorption ratio.

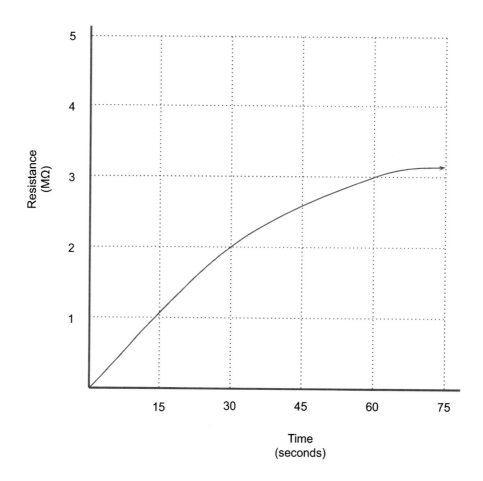

(A) The insulation has failed.

(B) The insulation is in questionable condition.

(C) The Insulation is in good condition.

(D) The insulation is in excellent condition.

23. A three-phase 150 MW steam driven synchronous generator at a local electric utility that was previously off line is slowly brought up to speed in order to supply power to the electrical grid. Select the protection relay below that is responsible for comparing the difference between the voltage phase angles and rotation sequence of the generator to the electrical grid during startup before allowing the generator breaker to close.

    (A)   ANSI #81

    (B)   ANSI #25

    (C)   ANSI #27

    (D)   ANSI #59

24. A three-phase, 150 HP, 460 V rated induction motor is used to operate a long conveyor belt that is loaded with heavy debris. When the motor attempts to start the conveyor belt from rest when the belt is fully loaded, the existing motor controller trips on overload. Select the next best motor starting application from the choices below to prevent overload conditions if speed control is not required. Assume that the motor is properly sized for the application.

    (A)   Soft Start Controller

    (B)   Variable Frequency Drive (VFD)

    (C)   Across the Line Starting

    (D)   Remote Starting

25. A PLC wiring diagram and the associated ladder logic are shown below. Describe what happens when switch 1 is toggled to the first throw position then toggled back to the off position it is currently shown in.

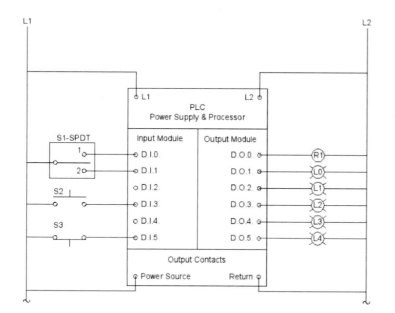

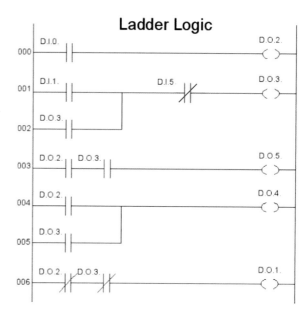

(A)   Relay R1 becomes energized and stays on.

(B)   Light L1 turns on and back off.

(C)   Light L1, light L4, and light L0 turn on and stay on.

(D)   Light L1 and light L3 turn on and back off, light L0 turns off and back on.

26. Calculate the minimum zone 2 line current setting to the nearest amp for a 0.82 lagging power factor load for the impedance relay protecting the three-phase 230 kV transmission line shown below if zone 2 is set for 75% reach and the positive sequence line impedance is 20 + j90 Ohms.

(A) 1,042

(B) 2,583

(C) 3,821

(D) 4,475

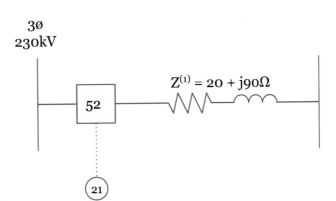

27. A three-phase wye connected 7.5 MVA, 4.160 kV synchronous motor has the following reactance values: X = 5 Ω, X' = 2.5 Ω, and X" = 1 Ω. Select the most accurate answer below that describes the excitation of the machine if it draws 922.5 amps at unity power factor and normal conditions.

  (A) The machine is over excited.

  (B) The machine is under excited.

  (C) The machine is neither over excited or under excited.

  (D) The excitation of the machine cannot be determined.

28. Select the answer below that is permitted according to the *2017 National Electrical Code®* for a single-phase, 60 Hz, 1 MVA, 69-4.16/2.402kV delta-wye connected dry type transformer:

  (A) The transformer may be installed indoors so long that it is located in a vault with a 3 hour fire resistance and concrete floors 4 inches thick.

  (B) The transformer may be installed indoors so long that it is located in a fire resistant room constructed with a minimum fire rating of 1 hour.

  (C) The transformer may be installed indoors so long that it is at least 12 inches from combustible material and located in a supervised location.

  (D) The transformer may not be installed indoors.

29. A single-phase, 60 Hz, 208 Vrms AC to 24 VDC power converter is used to charge a 24 Volt battery bank. Determine the minimum charge voltage that the battery bank must be rated for if the average DC voltage across the battery bank is 24.5V with an 11% ripple. Assume the ripple waveform is sinusoidal.

   (A) 24

   (B) 30

   (C) 32

   (D) 38

30. Select the most appropriate answer choice that satisfies the need of providing a total of 225 foot-candles for a large 50 foot by 60 foot conference room with a coefficient of utilization of 0.80 if energy consumption and color are not a factor.

   (A) 1,300 T2 fluorescent lamps at 540 lumens and 8 watts each.

   (B) 100 CFL flood bulbs at 225 lumens and 80 watts each.

   (C) 260 High pressure sodium lamps at 3250 lumens and 50 watts each.

   (D) 540 LED bulbs at 1250 lumens and 18.5 watts each (100 watt equivalent).

31. Select the answer below that will most likely result in a current transformer not being suitable for an overcurrent protective relay input device due to having a high percent of error if all other variables remain constant.

(A) A large increase in the number of turns on the primary CT winding.

(B) A large increase in the number of turns in the secondary CT winding.

(C) A large increase in the secondary CT excitation current.

(D) A large increase in the secondary CT current entering the connected device.

32. Calculate the percent impedance of the generator in the three-phase system shown below using the base values of the system. The system power base at the distribution bus is equal to the power rating of transformer T2, and the system voltage base at the distribution bus is equal to the secondary voltage rating of transformer T2.

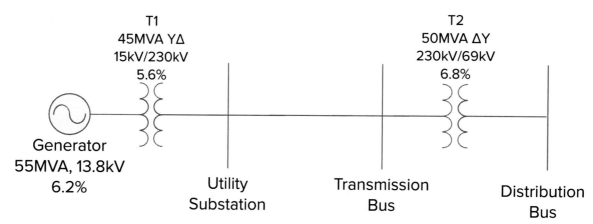

(A) 0.22%

(B) 4.8%

(C) 6.6%

(D) 9.2%

33. Select the only power supply and load configuration below that does not result in overloading the power supply while delivering the load rated voltage. Each power supply is balanced, serves no other load, and is capable of operating at full rated load. Assume ideal conditions with no voltage drop or power losses.

(A) A three-phase delta transformer secondary rated for 75 KVA and 208 volts providing power to a single-phase load rated for 75 KVA and 208 volts.

(B) A three-phase delta transformer secondary rated for 150 KVA and 240 volts with one center tapped winding providing power to a single-phase load rated for 75 KVA and 208 volts.

(C) A three-phase delta transformer secondary previously rated for 100 KVA and 208 volts with one open phase operating as an open delta providing power to a three-phase load rated for 75 KVA and 208 volts.

(D) None of the above.

34. Select the maximum standard size amperage rating of an instantaneous circuit breaker that may be used for the short circuit protection of the following induction motor according to the *2017 National Electrical Code®*. The motor nameplate is given below.

(A) 385

(B) 400

(C) 440

(D) 450

ELECTRIC MOTOR MANUFACTURER			
ORDER NO.		PQS 4817-135	
PHASE:	3	FRAME:	286T
HP:	30	SF:	1.15
FLC:	35	VOLTS:	460
RPM:	1780	HZ:	60
DESIGN:	B	AMB TEMP:	40 C
CODE:	G	ENCL:	TEFC
EFF %:	82	PF:	0.84

35. Determine the approximate capacitive reactance in ohms of a wye connected capacitor bank that results in a 15% voltage rise when shunted to the secondary side of a three-phase, 5 MVA, 13.8-4.16 kV delta wye transformer if the transformer has an impedance of 6.7%.

(A) 1.6

(B) 2.2

(C) 3.8

(D) 4.6

36. A worst case three-phase fault occurs downstream of a three-phase step down power transformer. Assuming that the transformer is fed by an infinite bus, select the answer below that best describes the available fault power in relation to the primary and secondary connections of the transformer.

(A) The available fault power at the primary connection of the transformer is greater than the fault duty of the infinite bus due to the power transformer let through characteristics.

(B) The available fault power at the secondary connection of the transformer is greater than the fault duty of the infinite bus since the fault occurs downstream of the transformer.

(C) The available fault power on the primary and secondary side of the transformer will be equal to the fault duty of the infinite bus.

(D) The available fault power on the primary and secondary side of the transformer will be equal to the ratio of the transformer's rated power and percent impedance when fed by an infinite bus.

37. A three-phase wye connected capacitor bank is connected to the customer bus in the single line diagram shown below. Select the answer that best describes the change in voltage at the customer bus due to the addition of the capacitor bank if the capacitor bank is rated for a capacitive reactance of 18 ohms per phase. The total wye connected load at the customer bus before the addition of the capacitor bank is 20 ohms at a lagging power factor of 0.92.

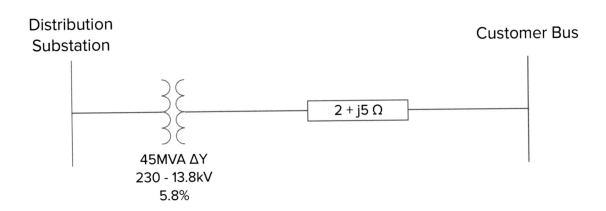

(A) The voltage at the customer bus has increased by 25%.

(B) The voltage at the customer bus has decreased by 25%.

(C) The voltage at the customer bus has increased by 15%.

(D) The voltage at the customer bus has not changed.

38. Select the answer below that will result in the greatest improvement to the overall power system stability of a complex electrical system made up of numerous synchronous machines that are interconnected through a network of transmission lines and power transformers.

    (A) Larger synchronous machine reactance values.

    (B) Larger transmission line series reactance values.

    (C) Larger power transformer leakage reactance values.

    (D) None of the above.

39. A 1000:5 multi-ratio CT is tapped for 600:5 resulting in a burden of 5.1VA. Determine the current on the line being monitored if the voltage drop across the CT due to excitation is 2.3 Volts.

    (A) 2.22 Amps

    (B) 222 Amps

    (C) 266 Amps

    (D) 300 Amps

40. According to the rate table below, determine the maximum demand power factor for the month for a medium power customer required to pay an equivalent total demand charge of $22,500 for the month if the maximum demand measured in kilowatts by the customer's demand meter during the same billing cycle is 3,500.

Demand Charge	$5.09 per kW of billing demand
First 100 hours of billing demand	4 cents/kWh
Next 50,000 kWh	2 cents/kWh
Remaining power	2.5 cents/kWh

*If power factor is less than 0.90, the billing demand will be calculated based off of the maximum KVA demand instead of the maximum KW demand.

(A) The medium power customer had a power factor greater than 0.70 but less than 0.80.

(B) The medium power customer had a power factor greater than 0.80 but less than 0.90.

(C) The medium power customer had a power factor greater than 0.90 but less than 1.

(D) The medium power customer had a power factor equal to unity.

Afternoon Session
Questions 41 - 80

41. According to the 2017 National Electrical Code®, select the minimum size copper AWG equipment grounding conductor required for a branch circuit that provides power to 208 volt, three-phase, combination non-motor load consisting of 15 kVA continuous and 5 kVA non-continuous. Assume that the minimum standard size overcurrent protection device is protecting the circuit.

(A) 14

(B) 12

(C) 10

(D) 8

42. A 3 amp load is connected to a 6 volt battery. Determine the internal resistance of the battery if the terminal output voltage is equal to 5.67 Volts.

(A) 0.11Ω

(B) 0.52Ω

(C) 0.81Ω

(D) 1.10Ω

43. A single-phase auto-transformer is used to step up a 200 kV bus. Calculate the power rating of the autotransformer to the nearest MVA if the power in the series winding is 6 MVA and the current delivered by the auto-transformer at full load and 0.82 power factor is 200 A.

(A) 22

(B) 34

(C) 46

(D) 58

44. Calculate the charging current in amps drawn by a three-phase 95 uF delta connected capacitor bank used for the power factor correction of a 60 Hz, 480 volt bus.

(A) 6

(B) 13

(C) 17

(D) 30

Electrical Power PE Practice Exam
7. Afternoon Session Questions

45. Which of the following most accurately describes the waveform of the smoothing capacitor instantaneous charging current supplied by the AC power source to an AC to DC rectifier.

   (A) The charging current will cycle continuously from positive to negative.

   (B) The charging current will cycle from positive to negative only when the internal capacitor voltage is greater than the voltage applied across the capacitor.

   (C) The charging current will cycle from positive to negative only when the voltage applied across the capacitor is greater than the internal capacitor voltage.

   (D) The charging current will be continuous and positive.

46. Determine the per unit reactive power rating of a capacitor bank required to correct the power factor of a 480 volt bus that supplies power to a pair of three-phase motors to 0.95 lagging if the first motor draws 0.5 pu of real power at a lagging power factor of 0.80 and the second motor draws 1 pu of apparent power at a lagging power factor of 0.72.

   (A) 0.4

   (B) 0.7

   (C) 0.9

   (D) 1.1

47. Determine the initial present worth value of a power transformer that was purchased 15 years ago if it has a salvage value of $18,000 and a declining balance rate of 0.04. The transformer depreciates the same amount of $5,000 each year and interest is 7%.

(A) $143,000

(B) $98,000

(C) $79,000

(D) $58,000

48. Select the most appropriate relay tap setting for the restraining windings in the ANSI #87 relay shown below for the differential protection of the three-phase transformer to reduce the amount of mismatch as much as possible if the primary CT ratio is 200:5 and the secondary CT ratio is 600:5.

(A) 5.3:5

(B) 7.2:5

(C) 8.5:5

(D) 10.5:5

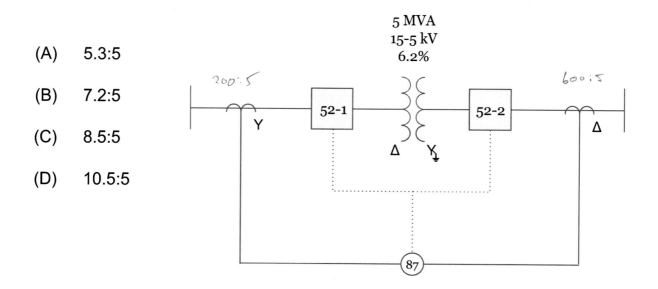

Electrical Power PE Practice Exam
7. Afternoon Session Questions

49. A 1,200 lumen pendant light is hung 2 feet from the ceiling in a 10 foot by 12 foot room with an overall height of 14 feet. Calculate the luminous intensity of the pendant light in candelas on a working plane 4 feet from the floor. Ignore the coefficient of utilization and assume there are no light loss factors.

    (A) 10

    (B) 64

    (C) 350

    (D) 640

50. A utility engineer is assigned to providing a service drop for a new industrial facility customer. Using the table data below, calculate the load factor of the new customer if the average power consumed is a constant amount of 20 MVA over the course of a month.

    (A) 1.17

    (B) 0.96

    (C) 0.86

    (D) 0.80

Load Description	Total Connected Load
120V non-motor loads	115kVA
480V non-motor loads	225kVA
Induction motor loads 200 Hp or larger	10MVA
Induction motor loads smaller than 200Hp	3MVA
All Synchronous motor loads	10MVA

51. Calculate the torque in lb·ft at rated conditions for a 60 Hz three-phase 150 HP motor that has two pole pairs and a slip of 0.08.

(A) 389

(B) 438

(C) 476

(D) 503

52. Four T8 fluorescent bulbs rated for 1025 lumens each are installed in a single light fixture to provide the light for a standard 10 foot by 12 foot office. The same fixture containing all four bulbs is relocated to a larger office and the light is measured. Select from the choices below the quantities that have most likely changed.

(A) The total luminance.

(B) The lamp burnout factor.

(C) The coefficient of utilization.

(D) All of the above.

53. A temporary receptacle is installed at a residential construction site to be used by on site qualified construction personnel only. Select the answer below that is in accordance with the *2017 National Electrical Code®*.

   (A) The temporary receptacle must have ground fault circuit interrupter protection provided by a ground fault circuit interrupter receptacle, a ground fault circuit interrupter circuit breaker, or a ground fault circuit interrupter cord plugged directly into the temporary receptacle.

   (B) The temporary receptacle must have ground fault circuit interrupter protection that may only be provided by a ground fault circuit interrupter receptacle or a ground fault circuit interrupter circuit breaker.

   (C) The temporary circuit is only required to have ground fault circuit interrupter protection if it is installed in a wet location.

   (D) The temporary circuit is not required to have ground fault circuit interruption protection as long as the temporary circuit is only used by qualified personnel.

54. Calculate the fault current in amps on a three-phase, 69 KV transmission line protected by an ANSI #51 relay with a current tap setting of 2.5 amps if the time dial setting of the relay causes it to operate at 5 multiples of pick up. The relay is connected to the line through a set of three wye connected current transformers with a ratio of 600:5 each.

   (A) 300

   (B) 600

   (C) 866

   (D) 1,500

55. A wye connected three-phase 55MVA, 13.8kV synchronous generator has an internal reactance equal to 1.4 pu. Calculate the internal stator voltage in kV at full load and a lagging 0.80 power factor.

(A) 17.2

(B) 18.2

(C) 19.1

(D) 20.0

56. Calculate the minimum ampacity rating in amperes for the disconnecting means of the runway conductors for an indoor bridge crane according to the *2017 National Electrical Code®* if the crane has one hoist motor, one trolley motor, and two bridge motors. The nameplate ratings of the motors are given below.

<u>Nameplate data</u>
Hoist motor: Three-phase, 60 Hz, 460 V, 75 HP, 92 A
Trolley motor: Three-phase, 60 Hz, 460 V, 25 HP, 31 A
Bridge motor: Three-phase, 60 Hz, 460 V, 20 HP, 24 A

(A) 108

(B) 116

(C) 125

(D) 142

57. Calculate the approximate percent voltage regulation of a three-phase synchronous generator operating at full rated load and a leading power factor of 0.83. The machine has a synchronous reactance of 0.04 per unit. Assume the per unit base values are equal to the ratings of the machine.

(A) -2.0

(B) 2.2

(C) 3.0

(D) -3.2

58. A single-phase 75 kVA rated 480/208V transformer is shorted on the low voltage side during a short circuit test. 52 volts applied on the primary side results in rated secondary current and 1,900 watts measured on the primary side. An open circuit test is then performed by energizing the secondary side with rated voltage while the primary is open. If 6.2 amps and 380 watts are measured on the secondary side during the open circuit test, calculate the transformer efficiency at full load and 0.82 power factor.

(A) 92%

(B) 94%

(C) 96%

(D) 98%

7. Afternoon Session Questions

59. Select the answer below that best describes the relationship between a voltage controlled PV bus from a complex power system modeled in a power flow software computational program.

(A) The real power and voltage magnitude of the bus are inputs, the reactive power and voltage angle of the bus in reference to the slack bus are calculated by the program.

(B) The real and reactive power of the bus are inputs, the voltage magnitude and voltage angle of the bus in reference to the slack bus are calculated by the program.

(C) The voltage magnitude and voltage angle of the bus in reference to the slack bus are inputs, the real and reactive power of the bus are calculated by the program.

(D) The reactive power and voltage angle of the bus in reference to the slack bus are inputs, the real power and voltage magnitude of the bus are calculated by the program.

60. A 480 volt balanced and positive sequence (ABC) three-phase power source is connected to an unbalanced three-phase load. The unbalanced load consumes 75 kVA of power with a lagging power factor of 0.97 in phase A, 65 kVA of power with a lagging power factor of 0.81 in phase B, and 45 kVA of power with a leading power factor of 0.72 in phase C. Calculate the impedance of a balanced wye connected load that consumes the same amount of three-phase power as the unbalanced load.

(A) 3.6 - j1.3 Ω

(B) 3.6 + j1.3 Ω

(C) 1.2 - j0.4 Ω

(D) 1.4 + j0.2 Ω

61. A locked rotor test is performed on a three-phase wye connected induction motor that has a 0.49 ohm leakage reactance and 6.67 ohm magnetizing reactance. Determine the total apparent locked rotor power to the nearest kVA if 200 amps of current and 96.5kW of real power are measured during the test.

   (A) 76

   (B) 95

   (C) 100

   (D) 113

62. Select the minimum standard trade size liquidtight flexible metal conduit in inches required according to the *2017 National Electrical Code®* for the following RHH conductors with the outer covering removed: three 4/0 AWG, and one 1/0 AWG.

   (A) 2

   (B) 2 1/2

   (C) 3

   (D) 3 1/2

63. A three-phase 480V induction motor draws an average current equal to 250 amps on each phase. Out of the choices below, select the most likely CT ratio for the current transformer used to meter the amperage drawn by the motor if the measured current on the secondary of the CT terminals is equal to 4.25 amps.

(A) 150:5

(B) 200:5

(C) 275:5

(D) 300:5

64. During insulation testing, a specific voltage is applied while the resulting current is monitored and recorded. Which of the various currents recorded is most likely to start out with the highest amperage magnitude while decaying the quickest?

(A) Capacitive Charging Current

(B) Dielectric Absorption Current

(C) Leakage Conduction Current

(D) Absorption Conduction Current

65. Select the minimum working space clearance in feet for an energized conductor that is 277 volts to ground with exposed energized parts on one side and grounded parts on the other side according to the *2017 National Electrical Safety Code®*.

   (A)   3 ½

   (B)   3

   (C)   2 1/2

   (D)   277 volts to ground does not require a minimum working space clearance.

66. The line voltage measured across a balanced wye connected load is 4,160 volts. Determine the complex load impedance if the load consumes 1,500 kVA of power at a lagging power factor of 0.92. Assume positive (ABC) sequence.

   (A)   32 - j13

   (B)   32 + j13

   (C)   11 - j5

   (D)   11 + j5

67. Select the answer below that best describes the heat losses in the copper windings of a transformer operating at 25% above rated nameplate conditions to the same transformer operating at full load.

(A) Heat losses will be approximately 125% compared to full load.

(B) Heat losses will be approximately 150% compared to full load.

(C) Heat losses will be approximately 175% compared to full load.

(D) There will be no change to the heat losses.

68. A current transformer with a 5 amp rated secondary winding is used as the relay input for 250% secondary overcurrent protection of a three-phase, 2 MVA, 230-4.16/2.40 kV, delta wye transformer that has an impedance of 5.6%. Select the most appropriate primary winding rating of the CT in amps.

(A) 300

(B) 450

(C) 600

(D) 700

69. Determine the line voltage magnitude in kilo-volts of a three-phase transmission system if the sum of all sequence impedance components during a single line to ground fault is equal to 10+j7.2Ω. The zero sequence current component is 646.6A<-35.8°.

(A) 24.0

(B) 13.8

(C) 10.2

(D) 786

70. Choose the answer below that most nearly describes the relationship between the voltage drop across the internal synchronous reactance and stator current of a synchronous motor compared to that of a synchronous generator during normal operating conditions.

(A) The synchronous reactance voltage drop leads the stator current by 90 degrees for both a synchronous motor and synchronous generator.

(B) The synchronous reactance voltage drop lags the stator current by 90 degrees for both a synchronous motor and synchronous generator.

(C) The synchronous reactance voltage drop leads the stator current by 90 degrees for a synchronous motor and lags the stator current by 90 degrees for a synchronous generator.

(D) The synchronous reactance voltage drop lags the stator current by 90 degrees for a synchronous motor and leads the stator current by 90 degrees for a synchronous generator.

71. Select the device setting change that will most likely result in the proper selective coordination for the system and TCC graph and show.

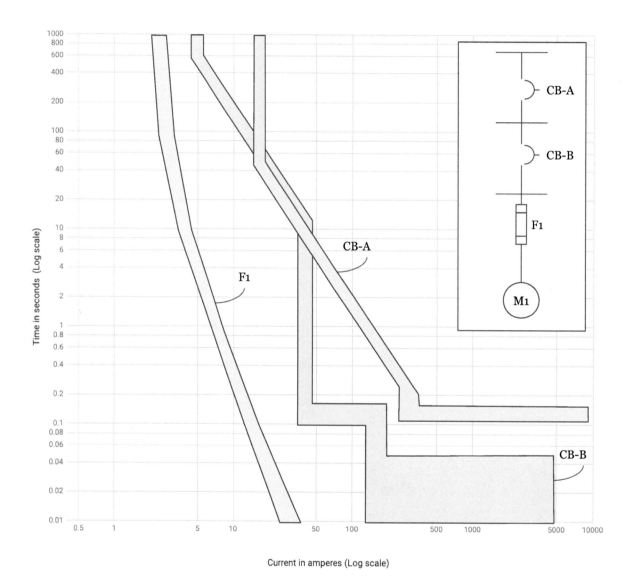

(A) Increase the long time tap setting for CB-A.

(B) Enabling instantaneous trip for CB-A.

(C) Increase the short time delay setting for CB-B.

(D) Decrease the long time tap setting for CB-A.

72. Select the statement below that is most accurate in regards to advantages of properly applied compensation techniques of transmission systems.

(A) Shunt compensation may be used to negate the harmful effects of harmonics.

(B) Shunt compensation may be used to absorb excess vars on the transmission line.

(C) Series compensation may be used to improve transmission power transfer.

(D) Series compensation may be used to increase voltage drop.

(This space intentionally left blank)

73. A variable frequency drive is used on a 60 Hz class B NEMA squirrel cage motor to slow it down to 50% of its rated frequency. Calculate the required voltage output in volts from the VFD that would maintain the same amount of flux in the motor present at rated operating conditions. The motor nameplate is below.

(A) 120

(B) 208

(C) 230

(D) 460

HP	30.00	SF	1.15
AMPS	34.9	VOLTS	3 PH 460
RPM	1767	HERTZ	60
CONT. DUTY	40C AMB.	MANF. DATE	1992
F CLASS INSUL	NEMA DESIGN: B	KVA CODE: G	NEMA NOM EFF: 93.6

74. Use the single line diagram shown below to select from the possible types of zones of protection that would still allow for Bus D to remain energized if a fault occurred in that zone.

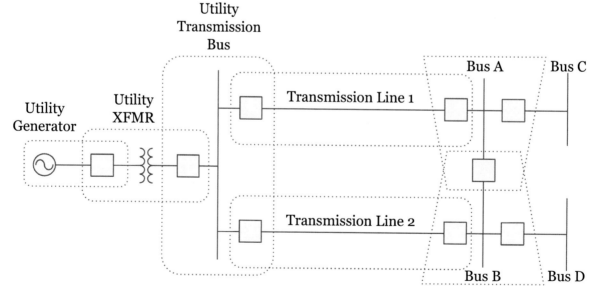

(A) Utility transmission bus protection

(B) Transmission Line 2 protection

(C) Utility transformer differential protection

(D) Bus B protection

75. Determine the minimum size service entrance conductor permitted according to the *2017 National Electrical Code®* for a small three-phase industrial facility with a maximum load of 375 amps if three sets of 75º C rated conductors are connected in parallel for each phase. Assume each set of phase conductors are run in separate raceways.

   (A)   2 AWG

   (B)   1 AWG

   (C)   1/0 AWG

   (D)   This application does not meet the requirements to parallel conductors and therefore is not permitted to code.

76. Select the answer below that best describes the block region setting of an ANSI #87 protective relay.

   (A)   The sensitivity of the protective relay is inversely proportional to changes in the block region.

   (B)   The sensitivity of the protective relay is not related to changes in the block region.

   (C)   The protective relay becomes less sensitive as the block region decreases.

   (D)   The protective relay becomes less sensitive as the block region increases.

77. Calculate the total inductive reactance in ohms for a three-phase, 230 kV, single conductor per phase, 60 Hz, 75 mile long transmission line if the phase conductors are horizontally spaced with the outer two phase conductors 5 feet from the center phase conductor. The diameter of each phase conductor is 2 inches.

(A) 11

(B) 21

(C) 42

(D) 83

(This space intentionally left blank)

78. 646 kVAR of reactive power is required to be supplied by a capacitor bank to a three-phase, 13.8 kV, 60 hertz system. Select the correct capacitor arrangement for each phase of the three-phase capacitor bank if it is balanced and delta connected. Each individual capacitor is rated for 2uF.

(A)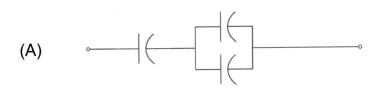

(B)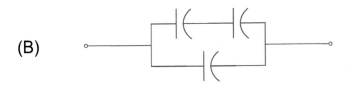

(C)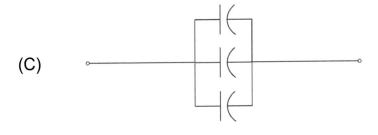

(D)

79. Select the answer below that is the most responsible reason that battery performance slowly deteriorates with the number of battery charge and discharge cycles even prior to reaching the rated cycle count.

- (A) Condensation build up inside the cell.
- (B) Cell wall deterioration.
- (C) Less reaction taking place.
- (D) Performance does not measurably deteriorate until the cycle rating is reached.

80. A single phase load with a complex impedance of 3+j2 ohms is connected across two terminals of a balanced three-phase 480V power source. Calculate the total amount of apparent power (kVA) drawn by the load.

- (A) 21
- (B) 64
- (C) 110
- (D) 192

Morning Session
Solutions 1 - 40

1. The answer is: (B) 10

We can calculate the C phase line current by applying the a operator to the b phase positive (1), negative (2), and zero (0) sequence current components accordingly and then adding them all together:

Positive Sequence (1) Components   Negative Sequence (2) Components   Zero Sequence (0) Components

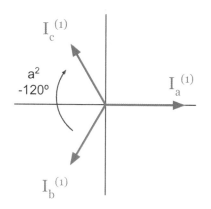 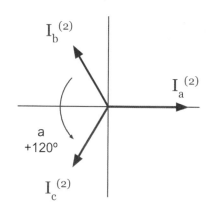 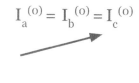

$$\hat{I}_c^{(1)} = a^2 \hat{I}_b^{(1)}$$
$$\hat{I}_c^{(1)} = 9.00A \angle -132° -120°$$
$$\hat{I}_c^{(1)} = 9.00A \angle -252°$$
$$\hat{I}_c^{(1)} = 9.00A \angle 108°$$

$$\hat{I}_c^{(2)} = a \hat{I}_b^{(2)}$$
$$\hat{I}_c^{(2)} = 3.75A \angle -135° +120°$$
$$\hat{I}_c^{(2)} = 3.75A \angle -15°$$

$$\hat{I}_c^{(0)} = \hat{I}_b^{(0)}$$
$$\hat{I}_c^{(0)} = 4.66A \angle 157°$$

The C phase line current is the sum of the C phase positive (1), negative (2), and zero (0) sequence current components:

$$\hat{I}_c = \hat{I}_c^{(1)} + \hat{I}_c^{(2)} + \hat{I}_c^{(0)}$$
$$\hat{I}_c = 9.00A \angle 108° + 3.75A \angle -15° + 4.66A \angle 157°$$
$$\hat{I}_c = 10.0A \angle 110°$$

The magnitude of the C phase line current to the nearest ampere is 10.

*Note: for $I_c^{(1)}$, a phase angle of -252° and 108° are equal. Either are fine to use in your calculations. Typically angles are kept between -180° and +180°. Try graphing $I_c^{(1)} = 9.00A \angle \mathbf{-252°}$ and $I_c^{(1)} = 9.00A \angle \mathbf{108°}$ to verify for yourself if this is new for you.

2. The answer is: (D) None of the above.

The NEC® has a list of all of the information that a motor shall (required to) be marked with in NEC® 430.7(A). These markings are typically stamped onto the motor nameplate. The first three possible answer choices are all required:

(A) Rated full load motor speed.

Required per NEC® 430.7(A)(4).

(B) Continuous or time rating.

Required per NEC® 430.7(A)(6).

(C) Rated heater voltage if applicable.

Required per NEC® 430.7(A)(15).

Since all of the possible answer choices **are** required, and none of the possible answer choices are **not** required, **the answer is none of the above.**

3. The answer is: (B) 102

A standard single-phase full wave rectifier allows the positive input peaks to pass through to the output, along with the rectified negative peaks by inverting them:

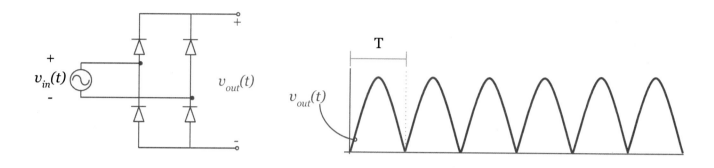

If IGBTs are used instead of the diodes shown in the circuit above, then the full wave rectified output voltage waveform will only be present when the gate of each IGBT is signaled between 0.926 ms to 7.41 ms for every period of the rectified voltage.

To simplify the coming integral, let's convert the firing angles of each IGBT from the time domain of seconds to degrees. Watch your units! 1 millisecond (ms) is equal to 0.001 seconds, or 1 ms = 1X10^{-3} s.

$$t = \frac{\theta}{360° \cdot f}$$

$$\theta_1 = 360°(60Hz)(0.926ms)$$
$$\theta_1 = 20°$$

$$\theta_2 = 360°(60Hz)(7.41ms)$$
$$\theta_2 = 160°$$

Now let's show one period (T) of the output voltage with the IGBT circuit:

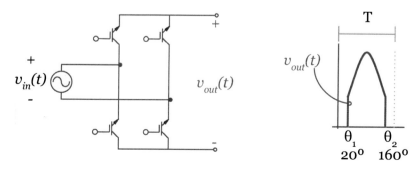

3. The answer is: (B) 102

(← continued from previous page)

Now it's time to integrate the sine function using the average value formula over the course of one period (T), using the firing angles (θ_2 and θ_1) as our upper and lower limits (b and a) of the integral. Since we are using the units of degrees for the firing angles, we'll also need to use the unit of degrees for the period (T). The period of a full wave rectifier is equal to 180 degrees, which is the duration of one peak of a sine wave (half of the 360 degree period of a standard sine wave).

You'll want to do this in your calculator using the integral button. Make sure your calculator's mode is set to degrees ("DEG" needs to be displayed in the top right hand corner of your calculator):

$$AVG = \frac{V_{peak}}{T} \int_a^b sin(t)dt$$

$$AVG = \frac{\sqrt{2}\,(120V)}{180°} \int_{20°}^{160°} sin(t)dt$$

$$AVG = 0.943(107.68V)$$

$$AVG = 102V$$

The average DC voltage output of the rectifier to the nearest volt is 102.

4. The answer is: (C) 4/0 AWG

The application of different temperature ratings of conductors according to code is commonly misunderstood.

According to temperature limitations *NEC® 110.14(C)*, conductor temperature ratings must be selected in order to "not exceed the lowest temperature rating of any connected termination, conductor, or device."

If a conductor with a greater temperature rating (for example, 90 C) is being used on equipment with a lower temperature rating (for example, 75 C), then ampacity of the conductor must be selected based on the conductor ampacity values at the lower temperature rating of the equipment (75 C), instead of the conductor ampacity values at the actual higher temperature rating of the conductor (90 C).

This may seem confusing at first. Let's work through this problem to have a better idea of how to put this into practice.

Step 1: The **FLC** of a three-phase, 208V, 60 HP motor according to *NEC® Table 430.250* is **169A**.

Step 2: According *NEC® 430.22*, conductors for a continuous duty motor must have a minimum ampacity of 125% of the motor FLC:

169A(125%) = **211.3A**

Step 3: The four possible answer choices are 90 C rated conductors according to the problem. We need to check the standard temperature rating of the motor according to the equipment provisions in the NEC®:

NEC® 110.14(C)(1)(b)(1) circuits rated for greater than 100A will be rated for 75 C (our circuit is rated for 211.3A).

Step 4: Even though all of the possible answer choices are conductors with 90 C rated insulation, we will have to select the conductor size based on the 75 C rated ampacities in the NEC® conductor ampacity tables since the circuit is rated for 75 C Conductors.

This is what the code means by *"conductors with higher temperature ratings, provided the ampacity of such conductors does not exceed the 75°C (167°F) ampacity of the conductor size used"* in *NEC® 110.14(C)(1)(b)(2)*.

4. The answer is: (C) 4/0 AWG

(← continued from previous page)

Step 4 Continued:

Now we may use *NEC® Table 310.15(B)(16)* to identify the correct size for a copper, 75 C (instead of 90 C due to the temperature provisions of the motor) rated conductor with a minimum ampacity of 211.3A.

211.3A is the minimum required ampacity of the conductor, which means as long as the conductor is rated for 211.3A, or greater, then it is applicable to code.

Since 211.3A does not correspond to an actual conductor size, and falls between copper 75 C 3/0 AWG (Rated for 200A) and copper 75 C 4/0 AWG rated for (230A), we must round up to 75 C 4/0 AWG.

4/0 AWG 75 C Copper has an ampacity of 230A, and is greater than the minimum required ampacity of 211.3A for this circuit.

The answer is 4/0 AWG.

Note 1: Answer (D) 250 kcmil would still be applicable to code since it has a greater ampacity rating than the minimum requirement of 211.3A for this circuit, however, it is still the incorrect answer for this problem since the question asks for the "minimum" size conductor, meaning the smallest possible size out of the answer choices that still meets code.

Note 2: If you ignored temperature provisions and sized the copper cable at the 90 C ampacity rating, you would have incorrectly chosen 3/0 AWG which at 90 C has an ampacity rating of 225A, but is undersized for this application when sized at 75 C (200A). Since the NEMA motor has conductor insulation ratings of 75 C, we may not use the 90 C ampacity of the cable, even though the conductor itself is rated for 90 C.

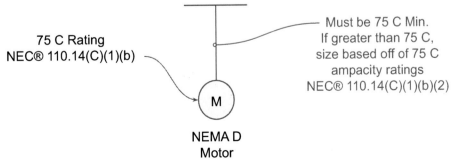

Electrical Power PE Practice Exam
8. Morning Session Solutions

5. The answer is: (D) Arc fault circuit interrupters protect residential circuits by de-energizing the branch circuit when a dangerous level of arcing is detected.

Arc fault circuit interrupters (AFCI) are designed to de-energize the branch circuit either by tripping the upstream branch circuit breaker or by opening the circuit directly at the receptacle when a **dangerous** level of arcing is detected.

They are not intended to de-energize the branch circuit when normal levels of arcing occur due to everyday use such as the spark that can sometimes be observed in receptacles when plugging or unplugging equipment, or when operating a standard 120 volt lighting switch.

AFCI protection comes in the form of two different methods. The first is an AFCI branch circuit breaker, installed directly into a residential panel that will protect the entire downstream circuit. The second is an AFCI receptacle that will protect the circuit at the point of the receptacle, and any non AFCI receptacles downstream that are daisy chained to the AFCI receptacle (similar to how a GFCI receptacle will protect all additional non GFCI receptacles that are powered from, and daisy chained to, the GFCI receptacle).

The main goal of an AFCI is to prevent fires that occur due to dangerous levels of arcing that may ignite nearby flammable material and is mostly required in residential bedrooms as applicable to code.

Let's evaluate each possible answer choice:

(A) Arc fault circuit interrupters protect residential circuits by de-energizing the branch circuit when the difference between the current on the hot and neutral conductor is greater than approximately zero amps.

This answer is describing a ground fault circuit interrupter (GFCI) and not an arc fault circuit interrupter (AFCI), **This is not the correct answer**. Ground fault circuit interrupters measure the current entering the circuit on the hot conductor, and leaving the circuit on the neutral conductor. During normal conditions, the current is equal ($I_H = I_N$) and the difference is zero ($I_H - I_N = 0$).

During a ground fault, the current drawn by the circuit on the hot conductor does not return on the neutral conductor since it is now shorted to ground by a ground fault.

When this occurs, the current on the hot and neutral conductor are no longer equal ($I_H \neq I_N$), and the difference between the current is no longer zero ($I_H - I_N \neq 0$).

5. The answer is: (D) Arc fault circuit interrupters protect residential circuits by de-energizing the branch circuit when a dangerous level of arcing is detected.

(← continued from previous page)

When this is detected, the GFCI will de-energize the circuit either at the breaker if the circuit is protected by a GFCI circuit breaker, or at the receptacle if the circuit is protected by a GFCI receptacle.

(B) Arc fault circuit interrupters protect residential circuits by de-energizing the branch circuit anytime a circuit overload occurs that could potentially lead to a fire.

While AFCIs are intended to prevent fires, they will de-energize the circuit during dangerous levels of arcing conditions, not overload conditions. **This is not the correct answer.**

(C) Arc fault circuit interrupters protect residential circuits by de-energizing the branch circuit when any amount of arcing is detected.

While an AFCI does de-energize the circuit due to arcing conditions, it only does so when a dangerous level is detected that could lead to a fire. By design, an AFCI is not intended to de-energize the circuit when normal and considered harmless levels of arcing occur such as plugging or unplugging equipment, or when operating a light switch. **This is not the correct answer.**

(D) Arc fault circuit interrupters protect residential circuits by de-energizing the branch circuit when a dangerous level of arcing is detected.

Arc fault circuit interrupters (AFCI) are designed to de-energize the branch circuit when a dangerous level of arcing is detected that could potentially lead to a fire.

This is the correct answer.

6. The answer is: (A) The circuit breaker will take more time to operate for the same level of fault current.

As the time characteristics of an overcurrent protection device becomes more time inverse, the negative slope of the time current curve (TCC) increases. This can be interpreted in two different ways.

From the perspective of fault current:

The overcurrent protection device will take longer to operate for the same level of fault current (see the vertical dashed line below).

From the perspective of time delay:

The overcurrent protection device will now require an increase in fault current to operate (see the horizontal dashed line below).

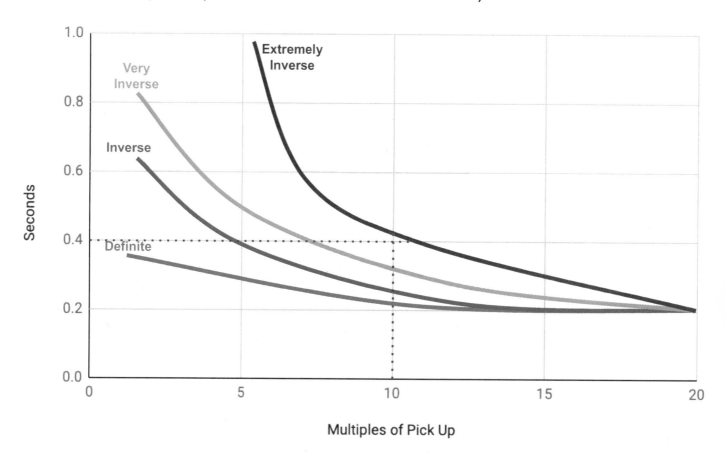

Notice that as the time characteristics are adjusted from *definite time* delay all the way to *extremely inverse time delay* without making changes to the pick up setting, the slope of the TCC increases sharply. The trip characteristics are becoming *less* sensitive to time.

6. The answer is: (A) The circuit breaker will take more time to operate for the same level of fault current.

(← continued from previous page)

Now let's evaluate each of the possible answer choices:

(A) The circuit breaker will take more time to operate for the same level of fault current.

By looking at the vertical dashed line at 10 multiples of pickup, we can see that by increasing the time inverse setting, the breaker will indeed take longer to operate for the same level of fault current. **This is the correct answer.**

(B) The circuit breaker will take less time to operate for the same level of fault current.

By looking at the vertical dashed line at 10 multiples of pickup, we can see that by increasing the time inverse setting, the breaker will take longer to operate for the same level of fault current, not less. This answer choice describes a change from a more inverse setting, to a less inverse setting. **This is not the correct answer.**

(C) The circuit breaker will operate at a lower level of fault current compared to the same time duration as before.

By looking at the horizontal dashed line at 0.4 seconds, we can see that by increasing the time inverse setting, the breaker will require a greater level of fault current to operate for the same duration of time as before. This answer choice describes a change from a more inverse setting, to a less inverse setting. **This is not the correct answer.**

(D) None of the above.

A *"none of the above"* answer choice can be difficult because it forces us to disprove all three other answer choices first in order to select it as the correct answer. Since one of the other three choices is the correct answer, **this is not the correct answer.**

7. The answer is: (C) 4 11/16 × 1 1/4 inch

This will be a lot easier with a quick sketch of the box:

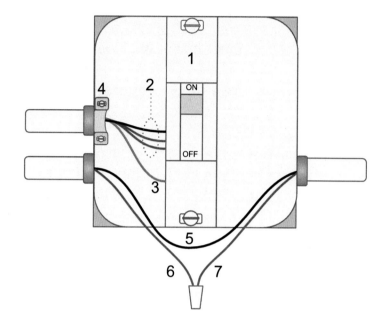

1. A yoke is the structural frame of a device, receptacle, or switch that is mounted to the box with screws. For conductor fill volume, each yoke counts as two conductors based on the largest conductor that is connected to it [NEC® 314.16(B)(4)]. Since size 12 AWG wires connect to the switch, **the switch counts as two 12 AWG conductors**.

2. Every conductor that originates outside of the box and terminates inside the box counts as one conductor based on its size [NEC® 314.16(B)(1)]. Three 12 AWG conductors from outside the box are run to the switch. **This counts as three 12 AWG conductors**.

3. Except for equipment grounding conductors that are used for isolated ground receptacles, only the largest equipment grounding conductor that enters the box is counted [NEC® 314.16(B)(5)]. There is one 12 AWG EGC in this box. **This counts as one 12 AWG conductor** (and would still only count as one 12 AWG conductor even if there were more than one EGCs entering the box).

4. Similar to EGCs, only one conductor volume allowance is given to clamps if there is one or more present in the box but this time it is based on the largest conductor inside the box [NEC® 314.16(B)(2)]. There is one clamp present and the largest conductor in the box is 12 AWG. **The clamp counts as one 12 AWG conductor** (and would still only count as one 12 AWG conductor even if there was more than one clamp present inside the box).

7. The answer is: (C) 4 11/16 × 1 1/4 inch

(← continued from previous page)

5. Every conductor that originates outside the box and passes through the box without being spliced (uncut) or termination, counts as one conductor. **The 12 AWG conductor that passes through the box counts as one 12 AWG conductor.**

6. Every conductor that originates outside the box and terminates or is spliced within the box counts as one conductor [NEC® 314.16(B)(1)]. **The 12 AWG conductor that is spliced (wire nutted) to the 14 AWG conductor counts as one 12 AWG conductor.**

7. Just like the previous step except now we are counting the smaller 14 AWG wire. **The 14 AWG conductor spliced (wire nutted) to the 12 AWG conductor counts as one 14 AWG conductor.**

Use NEC® Table 314.16(B) to count the total minimum fill volume of the box based on the conductor volume allowances from the 7 steps above:

1: 2 X 12 AWG Conductor = 2(2.25 in³)   switch
2: 3 X 12 AWG Conductor = 3(2.25 in³)   3 - #12
3: 1 X 12 AWG Conductor = 1(2.25 in³)   EGC
4: 1 X 12 AWG Conductor = 1(2.25 in³)   clamp
5: 1 X 12 AWG Conductor = 1(2.25 in³)   1 - #12 that passes through
6: 1 X 12 AWG Conductor = 1(2.25 in³)   1 - #12 spliced to #14
7: 1 X 14 AWG Conductor = 1(2.00 in³)   1 - #14 spliced to #12

The total minimum volume requirement of the box:

2(2.25 in³) + 3(2.25 in³) + 1(2.25 in³) + 1(2.25 in³) + 1(2.25 in³) + 1(2.25 in³) + 1(2.00 in³)

= 22.25 in³

Use NEC® Table 314.16(A) to look at the minimum volume size compared to the box dimensions given in the answers.

The smallest box out of the possible answers that is at least 22.25 in³ is the square metal 4 11/16 × 1 1/4 inch box that has a minimum volume of 25.5 in³ according to NEC® Table 314.16(A).

8. The answer is: (D) 879

First, draw the three phase delta - delta circuit:

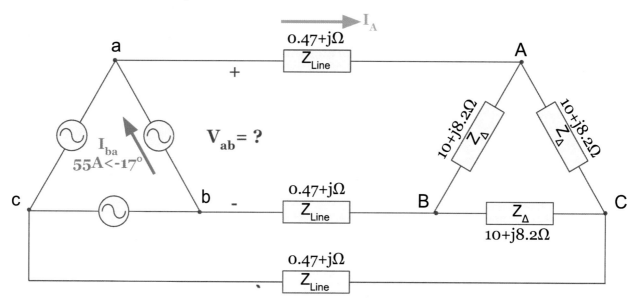

To determine the voltage of A phase delta source (V_{ab}), we can use a single-phase equivalent circuit to sum up the voltage using Kirchhoff's voltage law (KVL) and Ohm's law. We will need to calculate the A line current (I_A) and the wye equivalent impedance to use the single phase equivalent circuit:

$$\hat{I}_A = \sqrt{3}\,(55A) \angle (-17° - 30°) \qquad \hat{Z}_Y = \frac{10 + j8.2\,\Omega}{3}$$

$$\hat{I}_A = 95.3A \angle -47° \qquad \hat{Z}_Y = 3.33 + j2.73\,\Omega$$

Now let's draw the single-phase equivalent circuit and plug in these values. Remember that the single-phase equivalent circuit is just a model to help simplify our calculations for our three-phase circuit. Even though there is a neutral shown in the single-phase equivalent circuit, there is no neutral in our actual three-phase circuit since there is at least one delta connection:

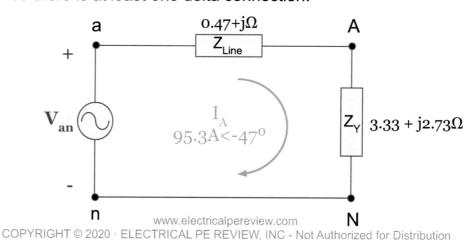

8. The answer is: (D) 879

(← continued from previous page)

Next, let's solve for the line to neutral voltage (V_{an}) of the single-phase equivalent circuit using Ohm's law:

$$\hat{V}_{an} = \hat{I}_A \cdot \hat{Z} = \hat{I}_A(\hat{Z}_{Line} + \hat{Z}_Y)$$

$$\hat{V}_{an} = (95.3A<\text{-}47°)(0.47+j\Omega + 3.33+j2.73\Omega)$$

$$\hat{V}_{an} = 507.4V<\text{-}2.5°$$

Last step is to convert the line to neutral voltage (V_{an}) of the single-phase equivalent circuit, to the line voltage (V_{ab}) of the system, which will equal the phase voltage of the delta connected power source:

$$\hat{V}_{ab} = \sqrt{3} \cdot |V_{an}|<(\theta_{Van} + 30°)$$

$$\hat{V}_{ab} = (\sqrt{3} \cdot 507.4V)<(\text{-}2.5° + 30°)$$

$$\hat{V}_{ab} = 878.8V<27.5°$$

The phase voltage magnitude of the delta connected power source to the nearest volt is 879.

9. The answer is: (B) 94%

To calculate the efficiency (η) of the motor we first need to determine both the real input power drawn by the motor from the bus (P_{in}) and the output power delivered by the motor to the mechanical load (P_{out}).

Even though the motor is rated for 460V, we will use 480V in our calculation for P_{in} since that is the actual voltage at the bus that is providing the 275 amps that the motor is drawing.

It helps to draw a diagram. All of the values used below are three-phase quantities:

$$P_{in} = \sqrt{3}\,|V_L|\cdot|I_L|\cdot PF$$
$$P_{in} = \sqrt{3}\,(480V)(275A)(0.87)$$
$$P_{in} = 198.9kW$$

$$P_{out} = 250Hp(746W)$$
$$P_{out} = 186.5kW$$

Now that we've determined both the input (P_{in}) and output (P_{out}) power, we can calculate the efficiency (η) of the motor:

$$\eta = \frac{P_{out}}{P_{in}}$$
$$\eta = \frac{186.5kW}{198.9kW}$$
$$\eta = 0.938$$
$$\eta = 93.8\%$$

The closest answer choice is 94%.

10. The answer is: (B) The poles in the rotor will lead the poles in the stator field by an increase of a factor of 3 times the previous mechanical angle.

Let's look at the formulas that relate mechanical angular displacement (α), electrical torque angle (δ), and internal voltage (Eo) to solve this qualitative question:

$$\delta = \frac{P \cdot \alpha}{2} \qquad \delta = \theta_{E_o} - \theta_E$$

If the internal voltage (Eo) begins to lead the terminal voltage (E) by a factor of three, the following change occurs to the electrical torque angle (δ):

$$\delta_{new} = 3 \cdot \theta_{Eo} - \theta_E$$

$$\delta_{new} = 3 \cdot \theta_{Eo} - 0°$$
$$\delta_{new} = 3 \cdot \theta_{Eo}$$
$$\delta_{new} = 3 \delta_{old}$$

Using a reference of zero degrees for the constant terminal voltage, we find that the electrical torque angle (δ) also increases by a factor of 3.

We can use this information to determine what change occurs to the mechanical torque angle (α) as a result:

$$\delta = \frac{P \cdot \alpha}{2}$$

$$\alpha_{old} = \frac{\delta_{old} \cdot 2}{P}$$

$$\alpha_{new} = \frac{(3 \cdot \delta_{old}) 2}{P}$$

$$\alpha_{new} = 3 \cdot \alpha_{old}$$

Since the mechanical torque angle (α) is linear to the electrical torque angle (δ), an increase by a factor of three to the electrical torque angle (δ) results in an increase by the same factor of three to the mechanical torque angle (α), also known as the mechanical angular displacement between the excited rotor and the rotating magnetic stator field in a synchronous machine.

11. The answer is: (D) Transformer inrush current occurs in the primary winding only and can potentially cause a differential protection relay to operate due to mismatch.

Let's evaluate all possible answer choices:

(A) Transformer inrush current is a sinusoidal component typically present only during transformer startup that is not seen by the differential protection relay due to being 180 degrees out of phase.

> Although transformer inrush current is typically only present during transformer startup it is not 180 degrees out of phase and it is seen by the differential relay. **Answer (A) is false.**

(B) Transformer inrush current does not contain a DC component and can potentially cause a differential protection relay to operate due to CT saturation.

> Transformer inrush current does contain a DC component, and although inrush current can in some cases lead to CT saturation and nuisance tripping of over current protection relays, it is not responsible for what may lead to nuisance tripping of differential protection relays. **Answer (B) is false.**

(C) Transformer inrush current is linear in both the primary and secondary winding and does not result in mismatch differential current.

> Transformer inrush current is not linear and it only occurs in the primary winding, because of this, it can lead to differential mismatch. **Answer (C) is false.**

(D) Transformer inrush current occurs in the primary winding only and can potentially cause a differential protection relay to operate due to mismatch.

> Since transformer inrush current only occurs in the primary winding, differential mismatch current occurs due to a larger increase in the transformer's primary CT relay current compared to the transformer's secondary CT relay current. This can cause nuisance differential tripping to occur.

> A common practice is to use a filter that blocks signals with non-fundamental frequencies from the relay to prevent inrush current and harmonics from causing a differential relay to trip when there is no internal fault present. **Answer (D) is true.**

12. The answer is: (D) The positive sequence voltage is equal to zero.

The positive sequence voltage component ($V_a^{(1)}$) is always zero during a three-phase fault, this question was designed to work you through proving this relationship. None of the given values mattered, they are just red herrings. Let's see why this is true.

Draw the three phase fault symmetrical component single phase equivalent circuit and fill in the given values:

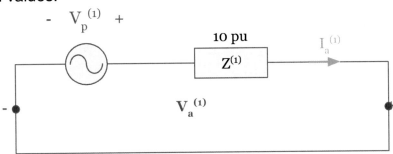

Now let's solve for the positive sequence voltage component ($V_a^{(1)}$) using (I) KVL, (II) Ohm's law, and (III) substitution to see why it equals zero:

(I) $\hat{V}_a^{(1)} = \hat{V}_p^{(1)} - \hat{I}_a^{(1)}(\hat{Z}^{(1)})$

(II) $\hat{I}_a^{(1)} = \dfrac{\hat{V}_p^{(1)}}{\hat{Z}^{(1)}}$

(III) $\hat{V}_a^{(1)} = \hat{V}_p^{(1)} - \left(\dfrac{\hat{V}_p^{(1)}}{\hat{Z}^{(1)}}\right)(\hat{Z}^{(1)})$

$\hat{V}_a^{(1)} = \hat{V}_p^{(1)} - \hat{V}_p^{(1)}$

$\hat{V}_a^{(1)} = 0$

The positive sequence voltage component ($V_a^{(1)}$) is equal to zero. It's even easier to see the relationship if we redraw the circuit as a standard single-phase equivalent circuit with the voltage source on the far left and a single load impedance on the far right:

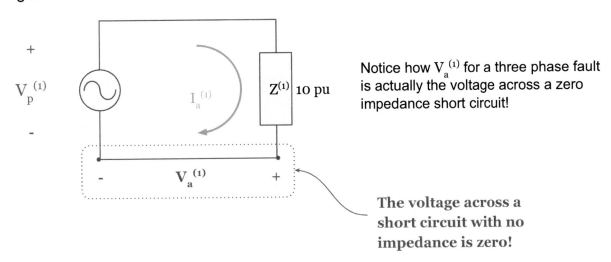

Notice how $V_a^{(1)}$ for a three phase fault is actually the voltage across a zero impedance short circuit!

The voltage across a short circuit with no impedance is zero!

13. The answer is: (C) Negative sequence harmonics oppose the rotating magnetic field in motors and can cause to motor windings to overheat as the motor draws more current.

The most common source of harmonics are switching mode power supplies, power electronics, and digital computer loads. Harmonics are created anytime a non-linear load is connected to a power system.

Just like symmetrical components, we can classify harmonics into three main categories:

Zero sequence harmonics (0) on the A, B, and C phase share the same angle of displacement. The result is that they sum at the neutral connection instead of cancelling out, (just like zero sequence components).

Example: The third (triplen) fundamental harmonic (f_3) is a **zero sequence harmonic**. Notice that the period is one third of the fundamental 60 Hz period and the phase angles are all equal:

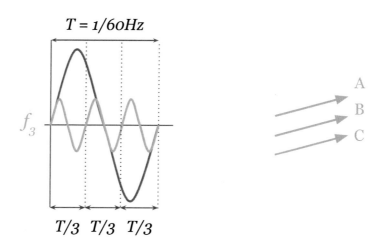

Harmful effects - Since zero sequence harmonics sum at the neutral, they can increase the amount of current the neutral conductor carries and can lead to overheating the neutral conductor.

Examples of zero sequence harmonics are the 3rd, 6th, 9th, 12th, 15th fundamentals.

The triplen harmonics, a subset of the zero sequence harmonics, are the 3rd, 9th, 15th, 21st, 27th fundamentals.

13. The answer is: (C) Negative sequence harmonics oppose the rotating magnetic field in motors and can cause to motor windings to overheat as the motor draws more current.

(← continued from previous page)

Positive sequence harmonics (+) on the A, B, and C phase are displaced by 120 degrees in the same ABC phase rotation sequence as the power system (just like positive sequence components). Since the positive sequence harmonics in each phase are displaced by 120 degrees, they cancel at the neutral.

Example: The fourth fundamental harmonic (f_4) is a **positive sequence harmonic**. Notice that the period is one fourth of the fundamental 60 Hz period and the phase angles are equally displaced by 120 degrees in the positive ABC sequence direction:

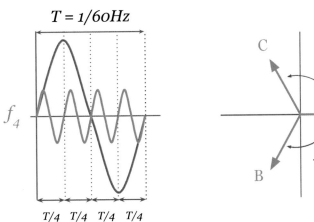
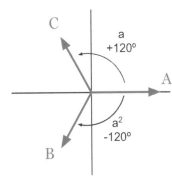

Harmful effects - Positive sequence harmonics circulate between phases and since they are in the same phase rotating sequence as the power system they are additive with the fundamental 60 Hz frequency. This can result in overheating phase conductors, transformers, motors, and any other current carrying device.

Examples of positive sequence harmonics are the 4th, 7th, 10th, 13th, 16th fundamentals.

13. The answer is: (C) Negative sequence harmonics oppose the rotating magnetic field in motors and can cause to motor windings to overheat as the motor draws more current.

(← continued from previous page)

Negative sequence harmonics (-) on the A, B, and C phase are displaced by 120 degrees in the opposite CBA phase rotation sequence compared to the power system (just like negative sequence components). Since the negative sequence harmonics in each phase are displaced by 120 degrees, they cancel at the neutral.

Example: The second fundamental harmonic (f_2) is a **negative sequence harmonic**. Notice that the period is one half of the fundamental 60 Hz period and the phase angles are equally displaced by 120 degrees in the negative CBA sequence direction:

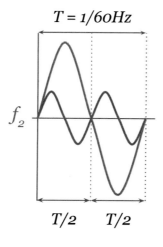
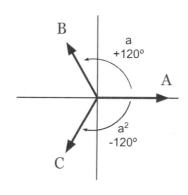

Harmful effects - Negative sequence harmonics circulate between phases and since they are in the opposite phase rotating sequence as the power system they oppose and counteract the rotating magnetic field of all rotating machines. This is because negative sequence harmonics are subtractive with the fundamental 60 Hz frequency. This can result in overheating of motors as the motor must draw more current to produce the magnetic field strength required to operate a mechanical load.

Examples of negative sequence harmonics are the 2nd, 5th, 8th, 11th, 14th fundamentals.

13. The answer is: (C) Negative sequence harmonics oppose the rotating magnetic field in motors and can cause to motor windings to overheat as the motor draws more current.

(← continued from previous page)

Now that we have a better understanding of the different types of harmonics, let's evaluate each possible answer:

(A) Negative sequence harmonics sum at the neutral point of three-phase wye connections and can lead to the neutral conductor overheating.

This describes the harmful effects of zero sequence (triplen) harmonics. This is not the correct answer.

(B) Negative sequence harmonics circulate between phases and can lead to overheating conductors due to their additive nature with the fundamental frequency.

This describes the harmful effects of positive sequence harmonics. This is not the correct answer.

(C) Negative sequence harmonics oppose the rotating magnetic field in motors and can cause to motor windings to overheat as the motor draws more current.

This correctly describes the harmful effects of negative harmonics. **This is the correct answer.**

(D) Unlike triplen harmonics, negative sequence harmonics are not harmful to the electrical system since they do not contribute to the total harmonic distortion of the system.

This is false, all harmonics contribute to the overall total harmonic distortion. This is not the correct answer.

14. The answer is: (C) 8

When a set of three CTs are delta connected to a three-phase system, there will be an additional √3 multiplier in the current seen by the relay. If a problem does not mention either in the question or on the diagram if a set of three CTs are wye or delta connected to a three-phase system, then always assume they are wye connected.

First let's calculate the short circuit current magnitude for 3 times the rated transformer current. Since the CTs and relay are located on the secondary side of the transformer, we'll have to use the transformer secondary rated amps:

$$|I_{Sec\ FLA}| = \frac{1,000kVA}{\sqrt{3}\ (4.16kV)} = 138.8A$$

$$|I_{Short\ Ckt}| = 3(138.8A) = 416.4A$$

416.4 amps of short circuit current is flowing into each of the three delta connected CTs on the secondary side of the transformer.

Next, let's use the CT ratio to calculate the secondary CT current. Since we are stepping down a line current value to a secondary CT current value, we will multiply the line current value by the CT ratio with the larger of the two CT ratio numbers on the bottom of the fraction:

$$|I_{CT}| = 416.4A\left(\frac{5}{450}\right) = 4.6A$$

4.6 amps of current is flowing out of each of the secondary CTs. Since the CTs are delta connected, we need to multiply this value by √3 to determine the current actually entering the relay. 4.6A(√3) = 8.0 Amps.

Let's draw a three-phase delta connected CT diagram in case you are not familiar with where the √3 multiplier comes from when dealing with the current leaving a delta connection.

The system is balanced and positive ABC sequence, so each of the three line currents and respective secondary CT currents will be displaced by 120 degrees from each other. For convenience, we will use a reference of 0 degrees for the A line current and the resulting phase angles for the remaining current in the following diagram and calculations:

Electrical Power PE Practice Exam
8. Morning Session Solutions

14. The answer is: (C) 8

(← continued from previous page)

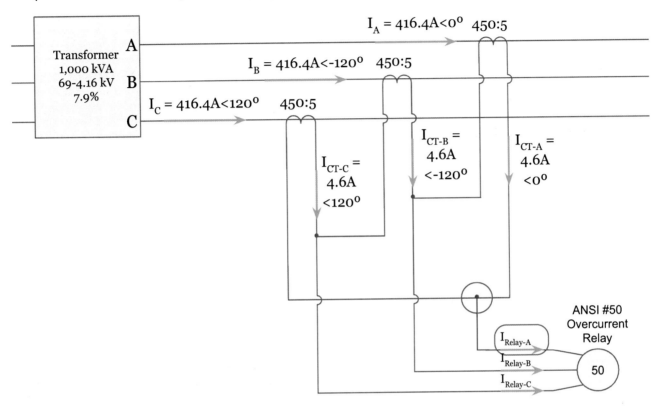

While the phase angle of each current entering the relay will be different, the magnitudes will all be the same. For convenience, we will solve for the first relay current circled in red, by using a KCL equation around the node also circled in red:

$$\hat{I}_{Relay-A} = \hat{I}_{CT-A} - \hat{I}_{CT-C}$$

$$\hat{I}_{Relay-A} = 4.6A<0° - 4.6A<120°$$

$$\hat{I}_{Relay-A} = 8.0A<-30°$$

$$|I_{Relay-A}| = 8.0A$$

The magnitude of current entering the relay is approximately 8 amps, or, √3 times 4.6 amps.

For more practice, try this problem again using a phase reference other than zero degrees for the A line current. The answer will still be the same. For additional help, consider reviewing connection diagrams and phasor diagrams for delta connected transformers.

15. The answer is: (A) 0.032 + j0.13

Transformer impedance can be referred to the primary or secondary side. To refer secondary impedance ($R_s + jX_s$) to the primary side, **multiply** the secondary impedance by the square of the transformer ratio (a^2). To refer primary impedance ($R_p + jX_p$) to the secondary side, **divide** the primary impedance by the square of the transformer ratio (a^2). Once the transformer impedance is referred to one side, sum the impedance together:

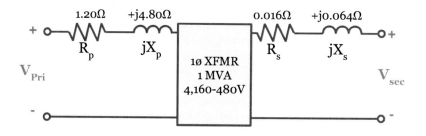

Transformer Equivalent Circuit

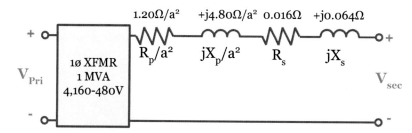

Transformer Equivalent Circuit Primary Impedance Referred to Secondary

(I) $\quad a = \dfrac{N_1}{N_2}$

$a^2 = \left(\dfrac{4{,}160V}{480V}\right)^2$

$a^2 = 75.11$

(II) $\quad \hat{Z}_{sec-eq} = \left(\dfrac{R_p + jX_p}{a^2}\right) + R_S + jX_S$

$\hat{Z}_{sec-eq} = \left(\dfrac{1.20 + j4.80\Omega}{75.11}\right) + 0.016 + j0.064\Omega$

$\hat{Z}_{sec-eq} = 0.032 + j0.13\Omega$

The total transformer impedance referred to the secondary side is **0.032 + j0.13Ω**.

16. The answer is: (B) 13.8-230 kV

Anytime three single-phase transformers are connected wye-delta or delta-wye, there is a "hidden" square root three in the three-phase voltage ratio. We can determine where the "hidden" square root three exists by drawing a three-phase transformer diagram.

Three single-phase 7.97-230 kV transformers connected wye-delta looks like this:

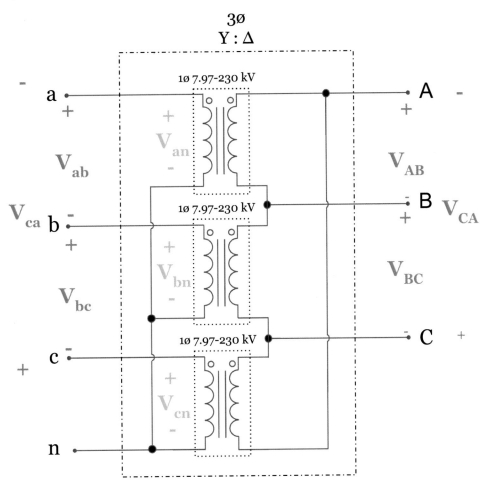

Pay attention to the difference between line voltages outside of the three-phase transformer boundary compared to the actual phase voltages across each single-phase transformer winding pair on both the primary and secondary side.

On the **primary wye side** of the **wye**-delta connected three-phase transformer, the line voltages (V_{ab}, V_{bc}, V_{ca}) are first reduced by a factor of $1/\sqrt{3}$ as they are converted from line to line voltages to line to neutral voltages (V_{an}, V_{bn}, V_{cn}), prior to being stepped up by each single-phase transformer by a single-phase transformer ratio of 7.97-230kV.

16. The answer is: (B) 13.8-230 kV

(← continued from previous page)

On the **secondary delta side** of the wye-delta connected three-phase transformer, the phase voltage across each pair of single-phase transformers is equal to the line voltage (V_{AB}, V_{BC}, V_{CA}) since there is no neutral and each single-phase transformer is connected directly to the line in delta configuration. Because of this, there is no additional "hidden" square root three on the secondary side.

This can also be demonstrated by looking at just one of the single-phase transformers that makes up the three-phase transformer:

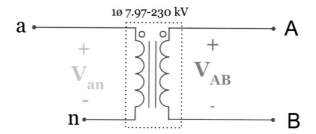

Notice how the single-phase transformer voltage ratio given in the problem is the primary and secondary voltage rating directly across each pair of single-phase transformer windings. Since the three-phase transformer is **wye-delta** connected, the primary side of each single-phase transformer is actually connected line to neutral (V_{an}) and the secondary side is connected line to line (V_{AB}).

Since the line voltage on the primary side of the **three-phase wye-delta transformer** is first reduced by a factor of $1/\sqrt{3}$, we can calculate the three-phase delta-wye voltage ratio by multiplying the primary voltage rating of the single-phase transformer by the square root of three:

$$V_{pri-3\phi} = \sqrt{3}\,(7.97kV)$$
$$V_{pri-3\phi} = 13.8kV$$

The three-phase wye-delta voltage ratio is then 13.8 - 230 kV.

Let's verify that a line voltage of 13.8 kV on the primary side of a wye-delta transformer made up of three single-phase 7.97-230 kV transformers would result in a line voltage of 230 kV on the secondary side. **Compare these values to their locations on the three-phase transformer diagram shown on the previous page:**

$$V_{ab} = 13.8kV \qquad V_{an} = \frac{13.8kV}{\sqrt{3}} = 7.97kV \qquad V_{AB} = 7.97kV\left(\frac{230kV}{7.97kV}\right) = 230kV$$

Looks good!

17. The answer is: (B) Awareness

The risk control method hierarchy is the order of control methods used to reduce risk ranked in the order of the most effective to the least effective:

1. **Elimination** - Remove the hazard altogether if possible.

2. **Substitution** - Complete the task with a different method or tool that is less hazardous than the current method.

3. **Engineering controls** - Redesign the equipment so that it is less hazardous.

4. **Awareness** - Bring more attention to the hazard by alerting personnel mostly through the use of visual aids and barricades.

5. **Administrative controls** - Improve the training and job planning to help decrease the hazard if possible.

6. **Personal protective equipment (PPE)** - Equipment such as hard hats, flame retardant clothing, and insulated tools to increase the protection of personnel.

Elimination should always be the first control implemented for a hazard if possible. If it is not possible, then the next highest control method should be implemented followed by the remaining control methods.

Personal protective equipment (PPE), is generally considered the "last resort" control method to protect personal as much as possible should they be exposed to the hazard.

The Hierarchy of Risk Control Methods can be found in the following location in the **Standard for Electrical Safety in the Workplace®: 2018 NFPA® 70E 110.1(H)(3)**.

18. The answer is: (D) $148

Since we are working in compounding periods (m) of months, we need to convert the nominal annual interest rate (r) to the effective interest rate (i) per compounding period. Since there are 12 months in one year, we simply divide r by 12:

$$i = \frac{r}{m} \qquad i = \frac{12\%}{12} = 1\%$$

let's draw the cash flow diagram for a) initial purchase agreement that has a monthly payment of $500 with no money down, and b) the alternate payment plan with $1,100 down and $125 a month:

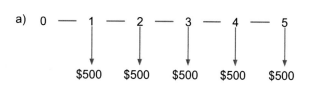
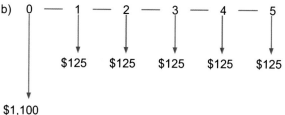

Next, let's calculate the equivalent monthly payment of the alternative payment plan by converting the $1,100 down to uniform monthly payments and then adding it to the existing $125 monthly payments:

$$A = A_1 + P(A/P, 1\%, 5)$$
$$A = \$125 + \$1,100\left(\frac{0.01(1+0.01)^5}{(1+0.01)^5 - 1}\right)$$
$$A = \$125 + \$1,100(0.2060)$$
$$A = \$351.60$$

Let's draw the equivalent cash flow diagram for the true uniform monthly payment amount of the alternative payment plan:

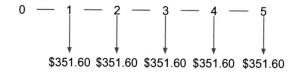

The last step is to compare the monthly payment cost of the original purchase plan to the alternate payment plan to determine the difference in cost between the two on a monthly basis: $500 - $351.60 = $148.40 per month.

The equivalent cost of $148 rounded to the nearest dollar per month is saved with the payment plan.

19. The answer is: (A) 0.89

Calculate the probabilities of a **parallel** paths by multiplying **unreliability (F)** values.
Calculate the probabilities of a **series** paths by multiplying **reliability (R)** values.

Let's draw the equivalent reliability block diagram (RBD) for the electrical system in this problem using the **unreliability (F)** values for parallel paths and **reliability (R)** values for series paths:

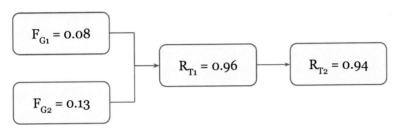

Now combine the **parallel** blocks by multiplying by the **unreliability (F)** values and combine the **series** blocks by multiplying by the **reliability (R)** values:

$$F_{G1 \& G2} = (0.08)(0.13) = 0.0104 \qquad R_{T1 \& T2} = (0.96)(0.94) = 0.9024$$

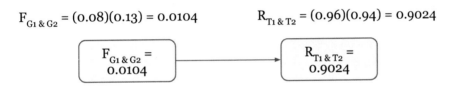

Now we are left with two series blocks. Before we can multiply these two together we need to convert all **unreliability (F)** values to **reliability (R)** values using the formula R + F = 1:

$$R_{G1 \& G2} + F_{G1 \& G2} = 1$$
$$R_{G1 \& G2} = 1 - F_{G1 \& G2}$$
$$R_{G1 \& G2} = 1 - 0.0104$$
$$R_{G1 \& G2} = 0.9896$$

We can now calculate the overall probability of reliability for the system by combining the last two series blocks by multiplying the **reliability (R)** values:

$$R_{system} = (R_{G1 \& G2})(R_{T1 \& T2}) = (0.9896)(0.9024) = 0.8930$$

The probability of reliable electric power to the customer bus is approximately 0.89, or 89%. When you get familiar with RBD diagrams, you can do the calculation all in one step, like this:

$$R = [1-(0.08)(0.13)](0.96)(0.94)$$
$$R = 0.8930$$

20. The answer is: (B) 0.4%

Use NEC® Chapter 9 Table 9 to look up the resistance and reactance for one 250 kcmil uncoated copper conductor in PVC conduit:

$R = 0.052 \ \Omega/1{,}000ft$
$X_L = 0.041 \ \Omega/1{,}000ft$

According to note 2 at the bottom of NEC® Chapter 9 Table 9, we cannot use the "Effective Z at 0.85 PF" values directly from the table since the power factor is not 0.85. Instead, we will have to calculate the new approximate effective impedance value for this conductor at a lagging power factor of 0.80 using the equation given in note 2:

$|Z_e| = R(PF) + X_L \cdot \sin[\cos^{-1}(PF)]$
$|Z_e| = 0.052(0.80) + 0.041 \cdot \sin[\cos^{-1}(0.80)]$
$|Z_e| = 0.0662 \ \Omega/1{,}000ft$

The effective impedance of one 250 kcmil uncoated copper conductor in PVC conduit (shown as "250 kcmil" below) at 0.80 lagging power factor is 0.0662 Ω/1,000ft.

Next we will need to calculate the impedance of each phase taking into consideration that there are three parallel conductors per phase:

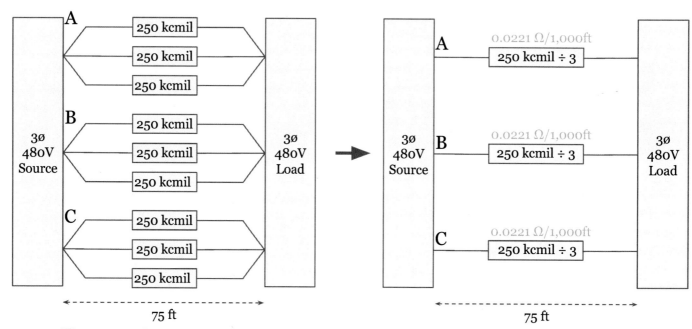

Three conductors run in parallel will have an equivalent impedance equal to 1/3rd of the original value. This means that each phase only sees 0.0662 ÷ 3 = **0.0221 Ω/1,000ft** of impedance per phase.

20. The answer is: (B) 0.4%

(← continued from previous page)

Now let's calculate the voltage drop across each phase conductor. Don't forget to use the circuit length to convert from ohms per 1,000 ft to ohms:

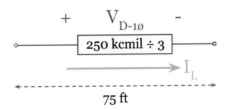

$$\hat{V}_{D-1\phi} = \hat{I}_L \hat{Z}_{Line}$$

$$\hat{V}_{D-1\phi} = (650A \angle -37°)\left(\frac{0.0221\Omega}{1,000ft}\right)(75ft) \qquad \theta = \cos^{-1}(0.80) = 37°$$

$$\hat{V}_{D-1\phi} = 1.08V \angle -37°$$

It's important to note that even though this is a three-phase system, the voltage drop across each phase is a single-phase, or "per phase" value. You'll notice we use the subscript of 1ø. This becomes important in the next step.

Voltage drop in percent ($V_{D\%}$) is the ratio of the magnitude of the voltage drop across the conductor, to the magnitude of the voltage supplied at the source.

We can calculate it two different ways. The first way (a) is to compare the per phase voltage drop to the per phase voltage at the source.

The second way (b) is to compare the line voltage drop to the line voltage supplied by the source. Just like a regular per phase voltage quantity, we can convert the magnitude to a line quantity by multiplying by √3.

a) Comparing phase values:

$$V_{D\%} = \frac{|V_{D-1\phi}|}{|V_p|} \cdot 100 = \frac{1.08V}{\frac{480V}{\sqrt{3}}} \cdot 100 = 0.39\%$$

b) Comparing line values:

$$V_{D\%} = \frac{|V_{D-3\phi}|}{|V_L|} \cdot 100 = \frac{\sqrt{3}(1.08V)}{480V} \cdot 100$$

$$V_{D\%} = 0.39\%$$

Notice that both methods results in the same percent voltage drop. **The closest answer is a voltage drop of 0.4%.**

21. The answer is: (A) 0.11 + j0.44

There are several different ways you can expect transformer ratios to be written on the PE exam. One of the more obscure ways appears in this problem. We can assume that 230 - 13.8Y/7.97 kV stands for a wye connected secondary with all voltages shown in the units of kV.

Since the impedance is located in the voltage zone on the secondary side of the transformer, we will use the transformer's secondary voltage as the base voltage in that zone. Since this is a three-phase system, we will be using the transformer's secondary line voltage (13.8 kV) instead of the transformer's secondary phase voltage (7.97 kV).

To calculate the line impedance in ohms, multiply the per unit impedance by the base impedance:

$$\hat{Z} = \hat{Z}_{pu} \cdot |Z_B|$$

$$\hat{Z} = (0.02 + j0.08 pu)\left(\frac{13.8 kV^2}{35 MVA}\right)$$

$$\hat{Z} = 0.11 + j0.44 \, \Omega$$

The impedance (Ω) of the line is 0.11 + j0.44.

22. The answer is: (C) The insulation is in good condition.

The dielectric absorption ratio is the ratio of insulation resistance measured at 60 seconds, to the insulation resistance measured at 30 seconds.

D.A.R. 60 to 30 seconds ratio Evaluation:

Less than 1.0	Failure Imminent
1.0 to 1.25	Questionable
1.4 to 1.60	Good
Greater than 1.60	Excellent

From the graph, approximately 3 megaohms were measured at 60 seconds and 2 mega-ohms were measured at 30 seconds:

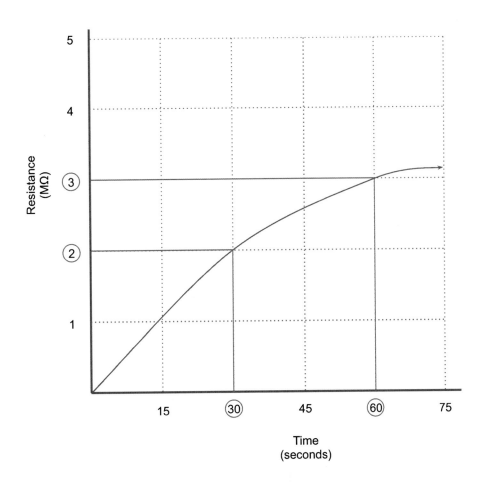

3MΩ/2MΩ = 1.5. The insulation is in good condition.

23. The answer is: (B) ANSI #25

The process of connecting a generator to an established electrical system is known as synchronizing.

One of the final permissive steps of synchronizing a generator to an established electrical system is ensuring that the generator and electrical system have the same phase sequence rotation (either positive ABC or negative CBA), and that the voltage being generated is in phase with the voltage of the system.

When the generator voltage and electrical system voltage are in phase, there is no angular phase displacement between the generator terminal voltage (E) on the A, B, and C phase compared to the electrical system.

The protective relay responsible for ensuring this is the ANSI #25 synchronizing relay.

24. The answer is: (A) Soft Start Controller

The motor does not have enough starting torque to get the conveyor belt moving If the existing motor circuit controller is tripping on overload conditions when the motor attempts to start the belt from rest after it is fully loaded with debris.

Let's evaluate each possible answer choice to see which one best suits this application:

A) Soft Start Controller

Soft start controllers are used to slowly ramp up a motors speed and torque using reduced voltage for a short period during the initial startup of the motor. Soft starters are best suited for applications that would benefit from speed and torque control only during initial starting and not during run time.

B) Variable Frequency Drive (VFD)

VFDs are used to modify a motor's speed, mostly for applications that require variable speed depending on changing mechanical loading conditions while running. While a VFD could be used for this motor to help with the starting torque and starting current, it would be more expensive than a soft starter and the problem states that speed control is not required.

C) Across the Line Starting

Across the line starting is the most common starting method for most induction motors applications. It consists of a heavy duty relay coil known as a motor starter that energizes the motor with full voltage as soon as the motor circuit is energized. According to the problem, this is most likely the method currently in place since the motor is tripping on overload during a high inertia starting event.

D) Remote Starting

Remote starting does not affect starting current or torque. Remote starting is just the use of hard wired push buttons to the control circuit possibly with the help of PLCs in order to energize the motor control circuit at a remote location that is further away from where the motor is located.

The best choice from the four possible answers given is a soft start controller.

25. The answer is: (D) Light L1 and light L3 turn on and back off, light L0 turns off and back on.

First let's trace out the logic and determine what is occurring at the default state shown with switch S1 in the "off" position:

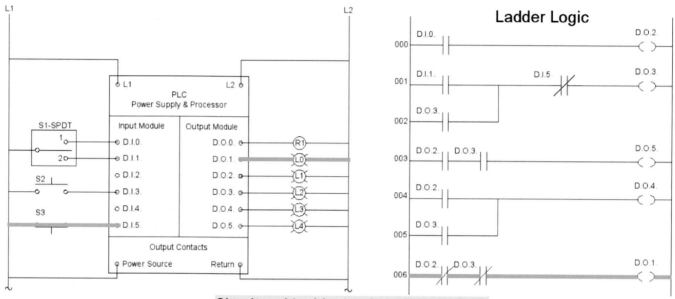

Circuit and Ladder Logic at Default State

In the default state, only light L0 is energized and turned on since digital output 1 (D.0.1.) on rung 006 is energized.

Now let's toggle switch S1 to the first throw position and see what changes:

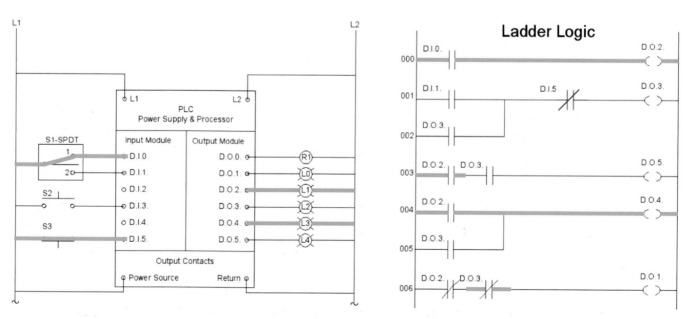

Circuit and Ladder Logic With S1 Toggled to First Throw Position

25. The answer is: (D) Light L1 and light L3 turn on and back off, light L0 turns off and back on.

(← continued from previous page)

With switch S1 now toggled to the first throw position, the digital input 0 (D.I.0.) on the PLC is now energized. This results in the following changes to occur inside the ladder logic in the following order:

1. <u>Rung 000</u>: digital input 0 bit (D.I.0.) changes from false to true.
    a. <u>Rung 000:</u> this changes digital output 2 bit (D.O.2.) from false to true.
        i. <u>Rung 003:</u> This changes digital output 2 bit (D.O.2.) from false to true.
        ii. <u>Rung 004:</u> this changes digital output 2 bit (D.O.2.) from false to true.
            - <u>Rung 004:</u> this changes digital output 4 bit (D.O.4.) from false to true.
    b. <u>Rung 006:</u> this changes digital output 2 bit (D.O.2.) from true to false.
        i. <u>Rung 006:</u> this changes digital output 1 bit (D.O.1.) 006 from true to false.

Due to the above changes in the ladder logic, the following changes occur to the PLC outputs:

1. With the digital output 2 bit (D.O.2.) on rung 000 now true (step 1.a. above), the associated digital output 2 (D.O.2.) on the PLC is now energized. Light L1 is connected to this digital output on the PLC and now becomes energized and turns on.

2. With the digital output 4 bit (D.O.4.) on rung 004 now true (step 1.a.ii. above), the associated digital output 4 (D.O.4.) on the PLC is now energized. Light L3 is connected to this digital output on the PLC and now becomes energized and turns on.

3. With the digital output 1 bit (D.O.1.) on rung 006 now false (step 1.b.i. above), the associated digital output 1 (D.O.1.) on the PLC is now de-energized. Light L0 is connected to this digital output on the PLC and now becomes de-energized and turns off.

Since there are no seal in bits in the ladder logic, as soon as switch S1 is switched back to the default off position, the circuit will revert back to the default state shown in the first diagram on the previous page with light L0 on, light L1 off, and light L3 off.

Turn to the next page for a summary of bit behavior based on switch operation →

25. The answer is: (D) Light L1 and light L3 turn on and back off, light L0 turns off and back on.

(← continued from previous page)

Summary of ladder logic bit behavior based on switch type and switch operation:

Type of PLC ladder logic input bit	Type of switch wired to the PLC input	Switch Operation	Resulting bit behavior
Examine if Closed (XIC) ─| |─	Normally Open (N.O.)	Normally Open	FALSE ─| |─
		Held Closed	TRUE ─| |─
	Normally Closed (N.C.)	Normally Closed	TRUE ─| |─
		Held Open	FALSE ─| |─
Examine if Open (XIO) ─|/|─	Normally Open (N.O.)	Normally Open	TRUE ─|/|─
		Held Closed	FALSE ─|/|─
	Normally Closed (N.C.)	Normally Closed	FALSE ─|/|─
		Held Open	TRUE ─|/|─

26. The answer is: (B) 2,583

First, let's draw the impedance phasor diagram:

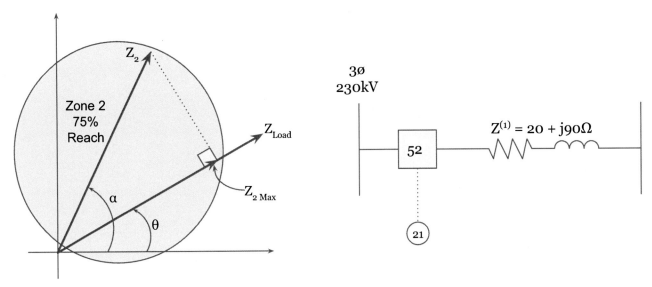

Next, let's calculate the complex impedance of the zone 2 phasor (Z_2).

Reach zone impedance phasors are always equal to the product of their percent reach and the positive sequence impedance ($Z^{(1)}$) of the line they are protecting:

$$\hat{Z}_2 = 75\% \hat{Z}_{line}^{(1)}$$
$$\hat{Z}_2 = 75\%(20+j90\Omega)$$
$$\hat{Z}_2 = 69.1\Omega < 77°$$

To calculate the minimum zone 2 line current that will result in a zone 2 trip, we will first need to calculate the maximum impedance value that will result in a zone 2 trip ($Z_{2\,Max}$).

The magnitude of $Z_{2\,Max}$ is equal to the real component of Z_2 projected along the complex load phasor (Z_{Load}):

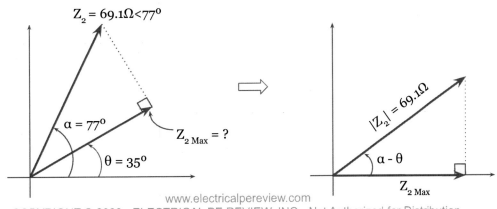

26. The answer is: (B) 2,583

(← continued from previous page)

$$|Z_{2\,Max}| = Re\{\hat{Z}_2\}$$
$$|Z_{2\,Max}| = |Z_2|\cos(\alpha-\theta)$$
$$|Z_{2\,Max}| = 69.1\Omega\cos(77°-35°)$$
$$|Z_{2\,Max}| = 51.4\Omega$$

$$\theta = \cos^{-1}(0.82)$$
$$\theta = 35°$$

Now that we know the maximum impedance value that will result in a zone 2 trip ($|Z_{2\,Max}|$), we can use the voltage of the transmission line to convert this to a line current value with Ohm's law.

Don't forget that when you use Ohm's law for a three-phase system, you must use the system's phase voltage value (line to neutral) and not the line voltage in order to solve for line current:

$$|V_p| = |I_L|\cdot|Z|$$

$$|I_L| = \frac{\frac{230kV}{\sqrt{3}}}{51.4\Omega}$$

$$|I_L| = 2,583A$$

A minimum line current of 2,583A will result in a zone 2 trip. Anything value less than this will not result in a zone 2 trip (it will be outside of the zone 2 trip circle in the impedance diagram).

27. The answer is: (C) The machine is neither over excited or under excited.

The excitation conditions of a synchronous machine can be determined by comparing the real component of the internal voltage (Eo) to the magnitude of the terminal voltage (E) when the terminal voltage is at a reference of zero degrees:

Re{Eo} > |E| The machine is over excited.
Re{Eo} = |E| The machine is neither over nor under excited (ideal excitation).
Re{Eo} < |E| The machine is under excited.

Synchronous machines that are over excited supply reactive power (Q) to the connected system.

Synchronous machines that are ideally excited neither supply nor consume reactive power (Q) to the connected system.

Synchronous machines that are under excited consume reactive power (Q) from the connected system.

Let's calculate both the terminal voltage (E) and internal voltage (Eo) using a reference angle of zero degrees for the terminal voltage. Since the machine is operating at normal conditions, we'll use the synchronous or "steady state" reactance value (X). Since the power factor is unity, the current angle will also be equal to zero:

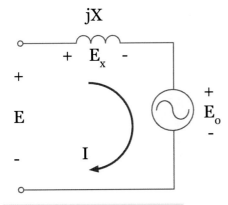

Synchronous Motor Single Phase Equivalent Circuit

$$|E| = \frac{4.160kV}{\sqrt{3}}$$

$$|E| = 2.402kV$$

$$\hat{E}_o = \hat{E} - \hat{I} \cdot jX$$

$$\hat{E}_o = 2.402kV - (922.5A)(j5\Omega)$$

$$\hat{E}_o = 5.200kV \angle -62.5°$$

Now let's draw a phasor diagram using a reference angle of zero degrees for the terminal voltage (E), and determine the excitation conditions of the machine:

27. The answer is: (C) The machine is neither over excited or under excited.

(← continued from previous page)

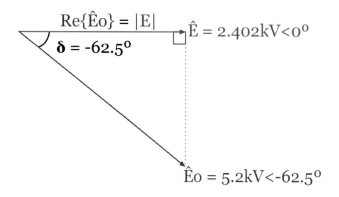

$$Re\{\hat{E}_o\} = |E_o|\cos(\delta)$$
$$Re\{\hat{E}_o\} = 5.200kV \cdot \cos(-62.5°)$$
$$Re\{\hat{E}_o\} = 2.401kV$$
$$Re\{\hat{E}_o\} \approx |E|$$

Since the real component of the internal voltage (E_o) is equal the magnitude of the terminal voltage (E), the machine is ideally excited (they are within one volt of each other with rounding). Note the right angle this creates between the two phasors in the phasor diagram.

Alternatively, you could have also used your calculator to convert E_o from polar to rectangular to determine the real component of E_o:

$$\hat{E}_o = 5.2kV \angle -62.5°$$
$$\hat{E}_o = 2.401kV - j4.612kV$$
$$Re\{E_o\} = 2.401kV$$

28. The answer is: (A) The transformer may be installed indoors in a vault built out of reinforced concrete 4 inches thick or greater.

NEC® 450.21 governs dry type transformers installed indoors.

According to *NEC® 450.21(C)*, since the transformer is over 35,000 volts (it has a 69 kV = 69,000V primary), it also must be installed in a vault that complies with *NEC® 450 Part III*.

Answer: the vault requirements of *NEC® 450.42* requires at least 3 hour fire resistance and concrete floors 4 inches thick or greater.

Even though the transformer is greater than 112 ½ KVA, the minimum fire rating of 1 hour from *NEC® 450.21(B)* (answer B) is not adequate since the transformer is also greater than 35,000 volts.

29. The answer is: (B) 30

The minimum charge voltage that the battery bank must be rated for needs to be at least equal to the maximum voltage of the output voltage $v_{out}(t)$ that the battery will see in this charging scheme. Let's start by drawing a diagram:

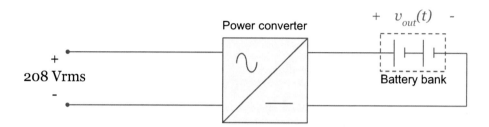

The average DC voltage across the battery is 24.5V, given by the problem. To find the maximum voltage that the output voltage $v_{out}(t)$ will reach at any given time can be determined by summing the average DC voltage and the peak voltage of the instantaneous ripple output voltage $v_r(t)$:

$$v_{out}(t) = VDC\,AVG + v_r(t)$$

First, let's calculate the RMS value of the instantaneous sinusoidal ripple voltage:

$$r = \frac{V_{r\text{-}rms}}{V_{DC}} \qquad V_{r\text{-}rms} = 0.11(24.5V)$$

$$V_{r\text{-}rms} = 2.695V$$

Next, let's use this value to calculate the peak voltage of the instantaneous sinusoidal ripple voltage:

29. The answer is: (B) 30

(← continued from previous page)

$$V_{r-peak} = \sqrt{2}V_{r-rms}$$
$$V_{r-peak} = \sqrt{2}(2.695V)$$
$$V_{r-peak} = 3.81V$$

Last, we can calculate the maximum value that the instantaneous output voltage will reach by summing the peak value of the instantaneous ripple voltage to the average DC voltage:

$$V_{Battery\ Max} = 24.5V + 3.81V$$
$$V_{Battery\ Max} = 28.3V$$

The battery has to at least be rated for a charge voltage of 28.3 volts:

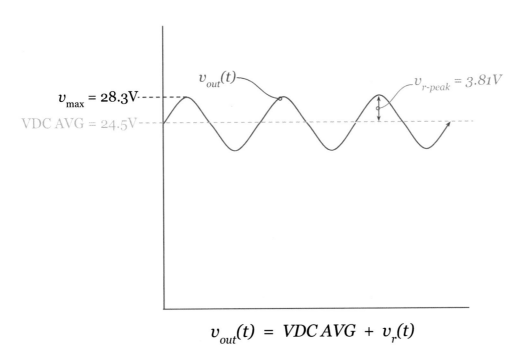

$$v_{out}(t) = VDC\ AVG + v_r(t)$$
$$v_{out}(t) = 24.5 + 3.81\sin(wt)$$

Out of the answer choices given, 30 is the minimum charge voltage that the battery must be rated for.

30. The answer is: (C) 260 High pressure sodium lamps at 3250 lumens and 50 watts each.

First, let's determine the total amount of lumens required to be provided in the room:

$$\Phi = E \cdot A$$
$$\Phi = (225 \ lm/ft^2) \cdot (50ft \cdot 60ft)$$
$$\Phi = 675,000 \ lm$$

Now let's factor in the coefficient of utilization and determine the total amount of lumens we would need our lighting system to provide:

$$\Phi = \frac{675,000 \ lm}{0.8}$$

$$\Phi = 843,750 \ lm$$

Now let's compare our answer choices.

Since color and energy consumption is not a factor, the type of light and wattage for each possible answer does not matter, only the total number of lamps and the lumen per lamp.

We will need to evaluate each possible answer based on the number of lamps (n) and lumens per lamp ($\Phi_{per\ fixture}$) to see which results in meeting the 843,750 lumen requirement.

The answer is C) 260 lamps at 3,250 lumens per lamp will result in the required amount of total lumens needed in the room:

$$\Phi = n \cdot \Phi_{Per\ fixture}$$
$$\Phi = (260 \ lamps)(3,250 \ lm \ per \ lamp)$$
$$\Phi = 845,000 \ lm$$

The other three answer choices result in a total lumen level less than the required amount.

31. The answer is: (C) A large increase in the secondary excitation current.

The formula for CT error is:

$$CT_{Error} = \frac{I_e}{I_{ct} + I_e} \cdot 100\%$$

I_e = The secondary CT excitation current.
I_{ct} = The secondary CT current entering the connected device.

Increasing the secondary CT current entering the device (I_{ct}) while keeping the secondary excitation current (I_e) constant will **decrease** the percent error.

Increasing the secondary excitation current (I_e) while keeping the secondary CT current entering the device (I_{ct}) constant will **increase** the percent error.

The answer is **(C) A large increase in the secondary excitation current.**

Try plugging numerical values in your calculator to verify this.

Increasing the number of turns on either the primary or secondary CT winding while holding all other variables constant will only change how much the CT is able to step down the primary line current.

32. The answer is: (B) 4.8%

There are three voltage zones in this problem. To correctly base change the generator per unit impedance, we will need to determine the base values in the voltage zone that the generator is located in.

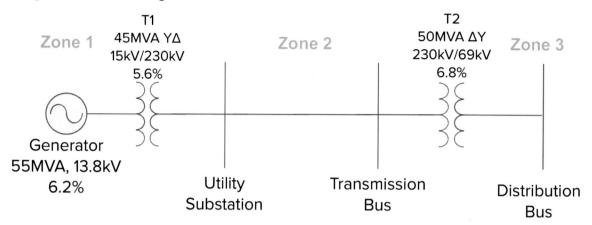

When you are working in per unit with a single line diagram, the system power base will be equal in each voltage zone but the system voltage base will be stepped up or down in each voltage zone according to each transformer ratio.

The power base in the Zone 1 where the generator is located, will be equal to 50 MVA, which is the same power base in Zone 3 at the distribution bus given in the problem statement.

To calculate the voltage base in the Zone 1, we will need to step the 69kV base voltage in Zone 3 up through transformer T2 from Zone 3 to Zone 2, then down through transformer T1 from Zone 2 to Zone 1:

$$V_B = 69kV \left(\frac{230kV}{69kV}\right)\left(\frac{15kV}{230kV}\right) = 15kV$$

The base voltage in Zone 1 is 15kV. Now we can base change the per unit impedance of the generator using the system base values in Zone 1, and the ratings of the generator as the old base values that the generator impedance is already expressed in:

$$Z_G = 6.2\% \left(\frac{Z_{B-old}}{Z_{B-new}}\right) = 6.2\% \left(\frac{\left(\frac{13.8kV^2}{55MVA}\right)}{\left(\frac{15kV^2}{50MVA}\right)}\right) = 0.0477$$

$$Z_G = 4.8\%$$

The percent impedance of the generator using the system base values is 4.8%.

33. The answer is: (B) 150 KVA 240V center tapped delta, 75 KVA 208V single-phase load

First, a quick note on voltage ratings of three-phase machines and three-phase systems before we get started. Any time a voltage rating of a three-phase machine or system is given, it's safe to assume that this is a line voltage unless otherwise stated. Typically, the ratings of a three-phase machine or system will be given in three-phase quantities (such as three-phase power, line voltage, and line current).

However, don't be confused if you are given the phase voltage rating of a delta configuration, as this will still be a line voltage value since the phase voltage of a delta connection is equal to the line voltage of the connected system.

Since the power supplies in choices A through C are all three-phase, the power (KVA) and voltage rating given for each are all three-phase quantities.

Let's begin by drawing a diagram for each power supply and load configuration to evaluate each of the possible answer choices to determine the best answer.

(A) A three-phase delta transformer secondary rated for 75 KVA and 208 volts providing power to a single-phase load rated for 75 KVA and 208 volts.

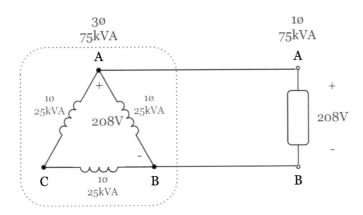

If the three-phase transformer with delta secondary is rated for 75 KVA, then each of the individual phase transformers that make up the three-phase transformer are rated for 75 KVA / 3 = 25 KVA. This means that the single-phase 75 KVA load will draw 3 times the amount of power from the phase it is connected to.

While the transformer is capable of providing the required load rated voltage of 208 volts, the phase windings of the transformer connected to the load will over load considerably if used for this application.

This is not the correct answer.

Electrical Power PE Practice Exam
8. Morning Session Solutions

33. The answer is: (B) 150 KVA 240V center tapped delta, 75 KVA 208V single-phase load

(← continued from previous page)

(B) A three-phase delta transformer secondary rated for 150 KVA and 240 volts with one center tapped winding providing power to a single-phase load rated for 75 KVA and 208 volts.

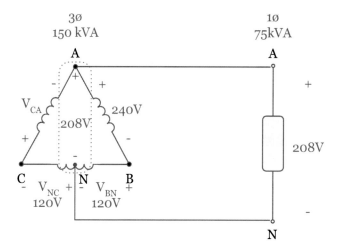

Even though the voltage of the transformer delta secondary is rated for 240 volts (VAB, VBC, VCA), if one of the windings is center tapped the transformer will be able to supply power at 208 volts single-phase.

Center tapping one of the windings (B phase above) will reduce the magnitude by half without changing the phase angle when measuring to the neutral from the phase that is center tapped. This means the B phase to neutral voltage magnitude is |VBN| = 240V / 2 = 120V.

If we use a reference of zero degrees for the A phase voltage (VAB) and assume positive ABC sequence, then the voltage magnitudes VBC (240V), VBN (120V), and VNC (120V), will all have an angle of negative 120 degrees, and the C phase voltage magnitude VCA (240V) will have an angle of positive 120 degrees.

Use your calculator in polar mode to verify that the magnitude of the voltage across the load (VAN) is approximately 208 volts. There are two ways to do this:

VAN = VAB + VBN   VAN = -VCA - VNC
VAN = 240V<0° + 120V<-120°   VAN = -240V<120° - 120V<-120°
VAN = 207.8V<-30°   VAN = 207.8V<-30°
|VAN| ≈ 208V |VAN| ≈ 208V

The power supply can supply the load rated voltage, but what about the power?

33. The answer is: (B) 150 KVA 240V center tapped delta, 75 KVA 208V single-phase load

(← continued from previous page)

Just like the previous possible choice (A), each individual phase transformer will have a power rating equal to one third of the three-phase transformer rating. The three-phase transformer with a delta center tapped secondary is rated for 150 KVA. Each individual phase transformer will be rated for 150 KVA / 3 = 50 KVA.

The first phase transformer winding from A to B is capable of supplying 50 KVA to the single-phase load.

The second phase transformer winding is capable of supplying half of its rated power to the single-phase load since the load is only connected to half of the number of turns at the center tap from B to N. This means that 50 KVA / 2 = 25 KVA worth of power can be supplied to the load from B to N.

The total amount of power that this power supply can provide to the single-phase load (assuming it serves no other loads according to the problem) is 50 KVA + 25 KVA = 75 KVA which is equal to the power rating of the connected single-phase load.

This power supply is capable of supplying the load rated voltage and power.

This is the correct answer.

(C) A three-phase delta transformer secondary previously rated for 100 KVA and 208 volts with one open phase operating as an open delta providing power to a three-phase load rated for 75 KVA and 208 volts.

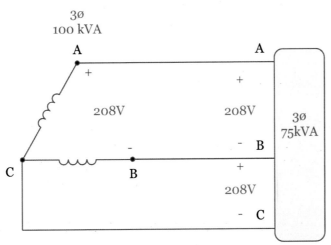

33. The answer is: (B) 150 KVA 240V center tapped delta, 75 KVA 208V single-phase load

(← continued from previous page)

When a three-phase delta transformer loses one phase and continues to operate as an open delta, it will continue to supply rated voltage to all three phases including the open phase (shown as the A phase above), however it will only be able to deliver 57.7% of the original power amount compared to the same three-phase delta transformer with all three phases intact.

Since we know the open delta can meet the three-phase load voltage rating of 208 volts across all three phases, we need to verify if it is large enough to deliver 75 KVA three-phase.

100 KVA · 57.7% = 57.7 KVA.

With one missing winding, the delta secondary now acting as an open delta, can only deliver a maximum of 57.7 KVA which falls short of the three-phase load requirement of 75 KVA. This would overload the transformer.

This is not the correct answer.

Note: It is worth noting that the 100 KVA delta secondary from answer (C) would have been adequate for the load if it still had all three phases intact prior to losing one phase and operating as an open delta.

For a more in depth look at the open delta transformer configuration, and to see where the 57.5% of power is derived from, please visit the following free article on the main www.electricalpereview.com webpage:

Electrical PE Review - Open Delta Transformer Connection:
https://www.electricalpereview.com/open-delta-transformer-connection/

34. The answer is: (D) 450

One of the biggest misunderstandings of applying the NEC® to motors is when to use the amperage of the motor that is marked on the nameplate, and when to use the full load current based on voltage and horsepower given in *NEC® tables 430.247-250*.

According to *NEC® 430.6(1)*, the full load current given in *NEC® tables 430.247-250* is to be used for sizing conductors, switches, short circuit protection, and ground fault protection for motors, **instead of** the full load amps marked directly on the motor name plate *unless* it is a motor that is less than 1200 RPM, high torque, or a multi-speed motor in which case the motor nameplate full load current is used for sizing.

In general, the only time you use the motor nameplate full load current is for sizing a separate motor overload device (see *NEC® 430.32(A)(1)* - "percent of motor nameplate full load current").

In this question, we are sizing the short circuit protection for a three-phase, 460 volt, 30 HP motor using an instantaneous trip circuit breaker.

Step 1: Use *NEC® Table 430.250* for three-phase alternating motors to look up the full load current to use for sizing according to the NEC® instead of using the full load current given in the name plate. According to the table, **the FLC is 40 Amps**.

Step 2: Use *NEC® Table 430.52* for the max rating of short circuit protection as a percentage of the full load current. According to the problem, we are sizing an instantaneous trip circuit breaker for a NEMA Design B (shown in the nameplate) motor.

According to the table, we are permitted to use a **maximum of 1,100% of the motor full load current**:

1,100%(40A) = 440A.

Step 3: *Exception No. 1* of *NEC® 430.52(C)(1)* in accordance with *NEC® Table 430.52*, instructs that if the value from Step 2 does not match a standard size, you may round up to a value that does not exceed the next standard size.

If you missed this exception in the code book, there was a clue left in the problem to select the maximum "standard size" amperage rating.

Standard sizes are listed in *NEC® 240.6(A)*.

34. The answer is: (D) 450

(← continued from previous page)

Step 4: 440 Amps does not match a standard size listed in NEC® 240.6(A), It falls between 400A and 450A. According to the exception, we are permitted to round up to a standard size of 450A as the maximum rating for an instantaneous trip circuit breaker used for the short circuit protection of this motor.

The answer is 450 amps.

If you incorrectly used the motor FLC, or incorrectly failed to round up to a standard size, you will choose the wrong answer.

If you need help with sizing overcurrent protection, conductors, or overloads for motors with different applications, look for the **Electrical PE Review Motor Cheat Sheet** typically given to students after our live class on motors.

35. The answer is: (A) 1.6

Let's start by drawing a diagram of the capacitor bank shunted to the secondary of the transformer. "Shunted" means connected in parallel:

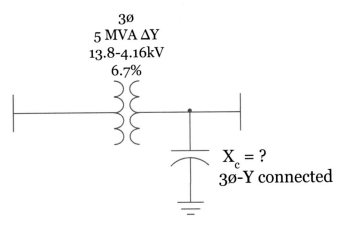

The percent voltage rise (or increase) of a transformer due to a shunted capacitor bank can be found using the following formula, where $S_{XFMR-SC}$ is the available short circuit power contribution of the transformer. Let's rearrange the terms to calculate the total amount of reactive power supplied by the three-phase wye connected capacitor bank:

$$V_{Percent\ Increase} = \frac{Q_c}{S_{XFMR-SC}} \cdot 100\% \qquad Q_c = 15\%\left(\frac{5MVA}{6.7\%}\right) = 11.19 MVAR$$

We can calculate the amount of reactive power being supplied by each phase of the capacitor bank by dividing by 3: $Q_{c\ 1\emptyset} = 11.19 \text{MVAR} / 3 = 3.73 \text{MVAR}.$

Let's draw the three-phase diagram of the wye connected capacitor bank:

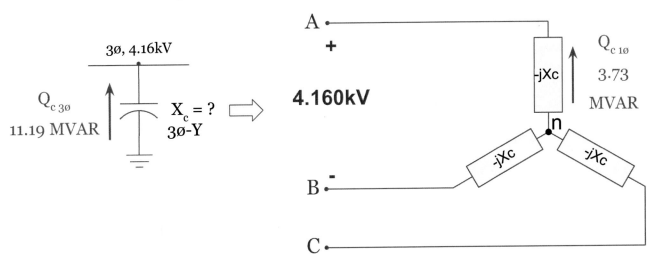

35. The answer is: (A) 1.6

(← continued from previous page)

We can now calculate the capacitive reactance (Xc) of the wye-connected capacitor bank. Notice the actual voltage across each phase of the capacitor bank in the previous diagram is the line to neutral voltage (VAn) since it is wye connected.

This means we will need to use the phase voltage (Vp) of the wye connection in our calculation:

$$X_c = \frac{|V_p|^2}{Q_{c1\phi}} = \frac{\left(\frac{4.160kV}{\sqrt{3}}\right)^2}{3.73MVAR} = 1.55\Omega$$

There is 1.55 ohms of capacitive reactance in each phase of the wye connected capacitor bank. The closest answer 1.6 ohms.

Don't forget, impedance (which can include resistance or reactance) is always a "per phase" quantity. It's always important to draw three-phase diagrams for delta and wye connections to help see which values are actually being applied to an impedance.

36. The answer is: (D) The available fault power on the primary and secondary side of the transformer will be equal to the ratio of the transformer's rated power and percent impedance when fed by an infinite bus.

When modeling an infinite bus during fault current conditions, the fault duty of the infinite bus is infinity. Let's use a very large number of 1,000,000 MVA to approximate the fault duty of the infinite bus, and a three-phase 1 MVA rated transformer with a 5% impedance to disprove choices A, B, and C:

(A) The available fault power at the primary connection of the transformer is greater than the fault duty of the infinite bus due to the power transformer let through characteristics.

(B) The available fault power at the secondary connection of the transformer is greater than the fault duty of the infinite bus since the fault occurs downstream of the transformer.

> The available fault power on the primary and secondary side of a transformer is always equal, it's the fault current magnitude that is either stepped up or down on either side of the transformer due to the transformer ratio. **(A) and (B) are false.**

(C) The available fault power on the primary and secondary side of the transformer will be equal to the fault duty of the infinite bus.

> Let's calculate the available fault duty on the primary and secondary side of the transformer using the hypothetical values from above:
>
> $S = (1 \times 10^6 \text{ MVA}) // (1 \text{ MVA} \div 5\%) = 20 \text{ MVA}$
>
> The available fault power on the primary and secondary side of the transformer is **not** equal to the 1,000,000 MVA fault duty of the infinite bus. **(C) is false.**

(D) The available fault power on the primary and secondary side of the transformer will be equal to the ratio of the transformer's rated power and percent impedance when fed by an infinite bus.

> The fault duty of the transformer is calculated by taking the ratio of the transformer's rated power and percent impedance: 1 MVA ÷ 5% = 20 MVA. This is equal to the available fault power we calculated in the previous step. This condition only exists when a transformer is fed by an infinite bus. **(D) is true.**

37. The answer is: (A) The voltage at the customer bus has increased by 25%.

To calculate the percent change in voltage with the addition of the capacitor bank, we'll need to calculate the voltage at the customer bus prior to the addition of the capacitor bank, calculate the voltage at the customer bus after the addition of the capacitor bank, then compare the difference.

First let's calculate the complex line current prior to the addition of the capacitor bank:

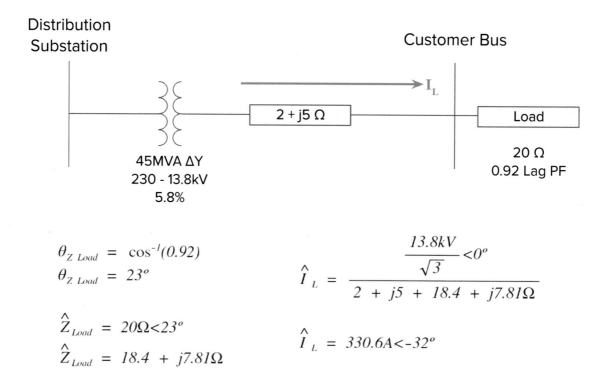

$\theta_{Z\ Load} = \cos^{-1}(0.92)$
$\theta_{Z\ Load} = 23°$

$\hat{Z}_{Load} = 20\Omega < 23°$
$\hat{Z}_{Load} = 18.4 + j7.81\Omega$

$\hat{I}_L = \dfrac{\dfrac{13.8kV}{\sqrt{3}} < 0°}{2 + j5 + 18.4 + j7.81\Omega}$

$\hat{I}_L = 330.6A < -32°$

Now let's use the line current to calculate the line to neutral voltage at the customer bus:

$\hat{V}_{Customer\ Bus\ LN} = \hat{I}_L \cdot \hat{Z}_{Load}$
$\hat{V}_{Customer\ Bus\ LN} = (330.6A < -32°)(20\Omega < 23°)$
$\hat{V}_{Customer\ Bus\ LN} = 6.612kV < -9°$

Next, let's add the capacitor bank to the diagram and calculate the new line current as a result of the new equivalent impedance:

37. The answer is: (A) The voltage at the customer bus has increased by 25%.

(← continued from previous page)

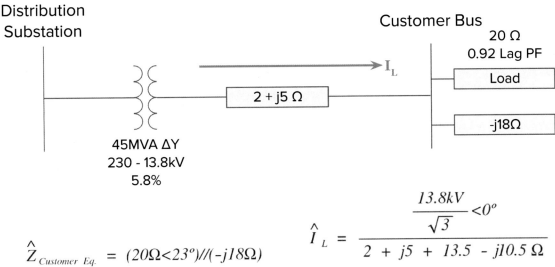

$$\hat{Z}_{Customer\ Eq.} = (20\Omega<23°)//(-j18\Omega)$$

$$\hat{Z}_{Customer\ Eq.} = 13.5 - j10.5\ \Omega$$

$$\hat{I}_L = \frac{\frac{13.8kV}{\sqrt{3}}<0°}{2 + j5 + 13.5 - j10.5\ \Omega}$$

$$\hat{I}_L = 484.4A<20°$$

With a change in both magnitude and angle for the line current, let's calculate the new line to neutral voltage at the customer bus using the new equivalent impedance at the customer bus:

$$\hat{V}_{Customer\ Bus\ LN} = \hat{I}_L \cdot \hat{Z}_{Customer\ Eq.}$$

$$\hat{V}_{Customer\ Bus\ LN} = (484.4A<20°)(13.5-j10.5\Omega)$$

$$\hat{V}_{Customer\ Bus\ LN} = 8.285kV<-18°$$

Last, we can compare the magnitude of the voltage available at the customer bus with the addition of the capacitor bank to what it was prior:

$$\frac{8.285kV}{6.612kV} = 1.25 = 125\%$$

The voltage at the customer bus is 125% of the original value prior to the addition of the capacitor bank, or we can say that **the voltage at the customer bus has increased by 25%**.

Note that we could have compared the line voltage magnitudes before and after the addition of the capacitor bank instead of the line to neutral voltage magnitudes and still arrive at the same answer.

38. The answer is: (D) None of the above.

Power system stability is the ability for synchronous machines, mostly generators providing power to the electrical system, to be able to return to a stable, steady state of operation after a sudden disturbance occurs on the electrical system such as a fault or sudden extreme changes in load.

The power transfer equation of a synchronous generator or two voltage buses connected by a transmission line can be expressed as:

$$P = \frac{|E_o| \cdot |E|}{X} sin(\delta)$$

$|E_o|$ = Magnitude of the generator internal voltage, or transmission line sending bus.
$|E|$ = Magnitude of the generator terminal voltage, or transmission line receiving bus.
X = Reactance of the machine, or transmission line.
δ = Phase angle difference between Eo and E.

Increased stability - Any change that results in an **increase** in the power transfer equation will always make a synchronous machine, or transmission system, more stable.

Decreased stability - Any change that results in a **decrease** in the power transfer equation will always make a synchronous machine, or transmission system, less stable.

(A) Larger synchronous machine reactance values.

Increasing the synchronous machine reactance (X) will result in a decrease in the power transfer (P) of the machine, resulting in less stability.

(B) Larger transmission line series reactance values & (C) Larger transformer leakage reactance values.

Increasing the transmission line reactance (X) will result in a decrease in the power transfer (P) of the transmission line, resulting in less stability. Increasing power transformer leakage reactance values will result in a greater equivalent transmission line reactance as seen by any connected generators.

As the equivalent line reactance of the transmission system as seen by a generator decreases, the greater the ability of each individual generator is capable of staying in synchronism with the other units by behaving as one large generator instead of many individual units.

(D) None of the above.

None of the possible answer choices result in an improvement to system stability. **This is the correct answer.**

39. The answer is: (C) 266 Amps

When a coil or inductor is energized it is known as being "excited."

Let's start by drawing what we know to get a better picture of the information given, and the information we need to solve for:

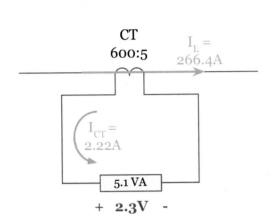

$$|S_{1\phi}| = |V_{1\phi}||I_{1\phi}|$$

$$I_{CT} = \frac{5.1\,VA}{2.3\,V}$$

$$I_{CT} = 2.22A$$

$$I_L = 2.22A\left(\frac{600}{5}\right)$$

$$I_L = 266.4A$$

Looking at the image it should be clear that we can determine the secondary current by plugging in the given burden and voltage drop into the formula for single phase apparent power magnitude.

From there we can use the ratio at the given tap position to find the line current. Since CT's are used to step down line current for metering, we know that the line current will be larger.

To make the secondary current larger we need to multiply by the CT ratio.

40. The answer is: (A) The medium power customer had a power factor greater than 0.70 but less than 0.80.

The total demand charge paid by the customer is $22,500.

We can use the demand charge rate from the billing table to determine the billing demand of the customer:

$$\frac{\$22,500}{\$5.09/kW} = 4,420.4 kW$$

This value is greater than the maximum demand measured by the demand meter for the month in kilowatts given in the problem (3,500KW). This means that the customer was not billed at their maximum kilowatt demand but by their maximum kVA demand due to poor power factor (see the power factor note at the bottom of the rate table in the problem).

Since we know that 4,420.4 is actually equal to the maximum kVA demand for the month from the customer, we can calculate the customer's maximum demand power factor:

$$P = S(P.F.)$$

$$P.F. = \frac{P}{S}$$

$$P.F. = \frac{3,500 kW}{4,420.4 kVA}$$

$$P.F. = 0.79$$

The customer's maximum demand power factor for the month is 0.79.

Since the customer had a power factor less than 0.90, the maximum KVA demand was used for their billing demand rate.

Afternoon Session

Solutions 41 - 80

41. The answer is: (D) 8

The minimum size for an equipment grounding conductor (EGC) is selected based off of the rating of the overcurrent device in the circuit according to NEC® Table 250.122, so first we must properly size the overcurrent device.

The minimum rating of the overcurrent device for continuous and non-continuous non-motor loads is the sum of 125% of the continuous loads and 100% of the non-continuous loads NEC® 210.20(A).

Let's calculate the minimum rating of the overcurrent device:

$$125\% \cdot 15kVA/(\sqrt{3} \cdot 208V) + 5kVA/(\sqrt{3} \cdot 208V) = 66A$$

Since according to code this is the *minimum* OCPD rating for the combination load, we will round up to the next standard size listed in NEC® Table 240.6(A) since the calculated version falls between two standard sizes.

The next standard size is 70 amps.

We can now look up the value of the OCPD in NEC® Table 250.122 to select the *minimum* EGC size. Notice not every standard size OCPD is listed in the table.

The table describes the OCPD sizes (ratting/setting) as *not exceeding*. The rating of the OCPD for this circuit (70A) falls between two sizes in the table, 60 and 100 amperes.

70A exceeds 60A, so the next table size up of 100A is used. The copper EGC size in the table for 100A is 8 AWG.

42. The answer is: (A) 0.11Ω

This is a simple single phase circuit analysis problem. You could ignore the word "battery" and still solve it correctly. Let's look:

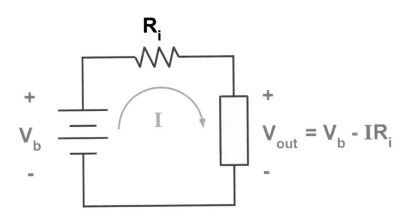

$$V_{out} = V_b - IR_i$$

$$R_i = \frac{V_b - V_{out}}{I}$$

$$R_i = \frac{6V - 5.67V}{3A}$$

$$R_i = 0.11\,\Omega$$

The battery has an internal resistance of 0.11 ohms.

43. The answer is: (C) 46

First, let's draw the auto-transformer with the information given:

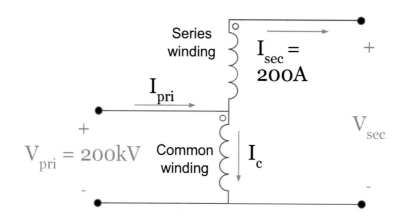

The power in the series winding (S_{ser}) and the power in the common winding (S_c) of an auto-transformer are always equal. Let's use this to set up a power formula for the power in the common winding (S_c).

The power in the common winding (S_c) is equal to the product of the voltage across the common winding (V_c) and the current through the common winding (I_c):

$$|S_c| = |V_c| \cdot |I_c|$$

Looking at the diagram above, the voltage across common winding (V_c) is equal to the primary voltage (V_{pri}), and the current through the common winding (I_c) is equal to the difference between the primary current (I_{pri}) and the secondary current (I_{sec}). Since there is no phase-shift between the primary and secondary of a single-phase auto-transformer, we can add or subtract magnitudes directly without having to worry about phase angles. The above formula now becomes:

$$|S_c| = |V_{pri}| \cdot (|I_{pri}| - |I_{sec}|)$$
$$|S_c| = |V_{pri}| \cdot |I_{pri}| - |V_{pri}| \cdot |I_{sec}|$$

The power rating of any transformer ($|S|$), including an auto-transformer, will always be the product of either the primary voltage (V_{pri}) and primary current (I_{pri}), or the product of the secondary voltage (V_{sec}), and secondary current (I_{sec}). Let's substitute and solve:

$$|S_c| = |S| - |V_{pri}| \cdot |I_{sec}|$$
$$|S| = 6MVA + (200kV)(200A)$$
$$|S| = 46MVA$$

The power rating of the auto-transformer is 46 MVA. Power factor is a red herring.

44. The answer is: (D) 30

Always draw diagrams for delta or wye connected three-phase devices to ensure that the proper values are used in three-phase formulas. Impedance is always a per phase quantity, so since the capacitor bank is delta connected we can assume that there are 95 uF in each phase of the capacitor bank:

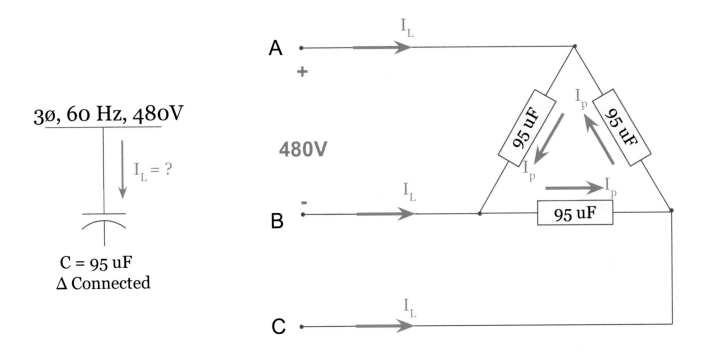

Charging current is the current drawn by a capacitor. Since there are no other loads on this circuit, the charging current will equal the line current drawn by the capacitor bank.

To calculate the line current, we will need to convert the capacitor values in farads to ohms first:

$$X_c = \frac{1}{2\pi f C} = \frac{1}{2\pi(60Hz)(95uF)} = 27.92\Omega$$

There are 27.92 ohms of capacitive reactance in each phase of the delta connected capacitor bank.

Next, let's use this information along with ohm's law to calculate the phase current of the delta connected capacitor bank. We'll use a reference of zero degrees for the voltage from A phase to B phase for convenience:

44. The answer is: (D) 30

($\leftarrow$ continued from previous page)

$$\hat{I}_p = \frac{\hat{V}_p}{\hat{Z}} = \frac{480V \angle 0°}{-j27.92\Omega} = 17.2A \angle 90°$$

The last step is to convert the phase current of the delta connected capacitor bank, to the line current drawn from the 480 volt bus. Since we are only interested in the magnitude of the line current, we'll drop the angle and only use the magnitude in our calculation:

$$|I_L| = \sqrt{3} \cdot |I_{P\triangle}|$$
$$|I_L| = \sqrt{3} \cdot (17.2A)$$
$$|I_L| = 29.8A$$

The charging current drawn by the three-phase delta connected capacitor bank is approximately 30 amps.

45. The answer is: (C) The charging current will cycle from positive to negative only when the voltage applied across the capacitor is greater than the internal capacitor voltage.

Smoothing capacitors are used in rectifier circuits to decrease ripple in the DC output by helping to resist changes in voltage variation.

Anytime the internal voltage of the capacitor is greater than the voltage applied to the capacitor from the rectifier, the capacitor will **discharge**. This occurs during **discharge cycles.**

Anytime the voltage applied to the capacitor from the rectifier is greater than the internal capacitor voltage, the capacitor will **charge** and draw current known as "charging current". This occurs during **charging cycles**:

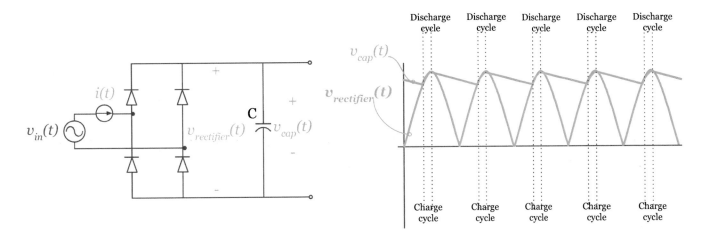

If the capacitor charging current is measured with an ammeter at the AC power source as the question indicates, then the charging current will cycle from positive to negative since the voltage reference at this point is also cycling from positive to negative:

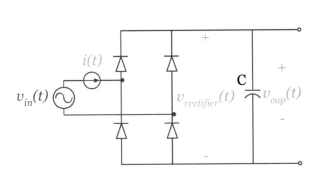

46. The answer is: (B) 0.7

First, let's draw the system:

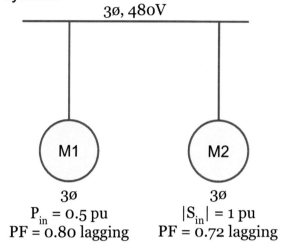

From the context clues in the problem, we know that the 480 volt bus must be a three-phase bus in order to provide power to three-phase loads. This means the capacitor bank will also be three-phase.

Per unit quantities are unitless, so we will have to derive which specific power quantities the problem gave for each motor using the keywords in the problem. Real power for motor one is in the units of watts when not expressed in per unit. To note this we will use the variable P. Apparent power for motor two is the magnitude of complex power in the unit of volt-amps when not expressed in per unit. To note this we will use the variable S.

Since both power quantities given are the amount of power each motor draws from the bus we know these are input power quantities (P_{in} and S_{in}) and not output power quantities (P_{out} and S_{out}). Had output power quantities been given, we would have also required each motor efficiency rating (η) to convert them to input quantities before proceeding with the problem ($P_{in} = P_{out}/\eta$ and $S_{in} = S_{out}/\eta$).

Let's begin by calculating the three-phase per unit complex power drawn by each motor from the 480 volt bus:

$$|S_{M1}| = \frac{P_{M1}}{PF_{M1}} = \frac{0.5 \; pu}{0.80} = 0.625 \; pu \qquad \theta_{M2} = \cos^{-1}(0.72) = 44°$$

$$\theta_{M1} = \cos^{-1}(0.80) = 37° \qquad \hat{S}_{M2} = 1pu\angle 44°$$

$$\hat{S}_{M1} = 0.625pu\angle 37°$$

46. The answer is: (B) 0.7

($\leftarrow$ continued from previous page)

Now let's calculate the total per unit complex power ($S_{3\phi}$) and power factor (PF) seen by the 480 volt bus prior to the addition of the capacitor bank by summing the per unit complex power drawn by each motor, and taking the cosine of the resulting power angle.

Don't forget to use your calculator to convert the per unit complex power seen by the 480 bus from polar to rectangular so that we also have the real ($P_{3\phi}$) and reactive ($Q_{3\phi}$) power quantities that will be needed to draw the power triangle of the 480 volt bus below:

$$\hat{S}_{3\phi} = \hat{S}_{M1} + \hat{S}_{M2}$$

$$\hat{S}_{3\phi} = 0.625pu<37° + 1pu<44°$$

$$\hat{S}_{3\phi} = 1.622pu<41°$$

$$\hat{S}_{3\phi} = 1.218 + j1.071 \ pu$$

$$PF = cos(41°)$$

$$PF = 0.75 \ lagging$$

Let's use the this information to draw the power triangle for the 480 volt bus prior to the addition of the capacitor bank:

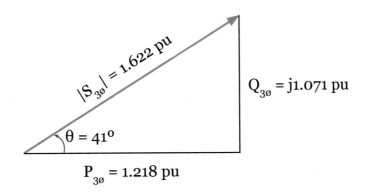

Next, let's draw the power triangle for the 480 volt bus after the addition of the capacitor bank that will correct the system power factor from 0.75 lagging to 0.95 lagging. To do this we need to calculate what the new power angle of the system will be at the new power factor:

$$\theta_{New} = cos^{-1}(0.95) = 18°$$

46. The answer is: (B) 0.7

(← continued from previous page)

When you correct (improve) the power factor of a system, the apparent power |S|, power angle (θ), and reactive power (Q) supplied by the system decreases. However, the real power (P) stays the same. This means that the amount of real power supplied by the bus after the addition of the capacitor bank will be the same as the amount of real power supplied by the bus prior to the addition of the capacitor bank ($P_{new} = P_{3\emptyset} = 1.218$ pu).

Now we are ready to draw the power triangle of the 480 volt bus after a capacitor bank has been added that will correct the overall power factor of the system to 0.95 lagging:

$\theta_{new} = 18°$, $Q_{New} = ?$, $P_{new} = 1.218$ pu

Let's use this information to solve for the amount of per unit reactive power supplied by the 480 volt bus after the addition of the capacitor bank:

$$Q_{New} = P_{New} \tan(\theta) = 1.218 pu \cdot \tan(18°) = 0.396 \ pu$$

We now have enough information to compare the power triangle of the system prior to and after the addition of the power factor correcting capacitor bank. It helps to draw them side by side:

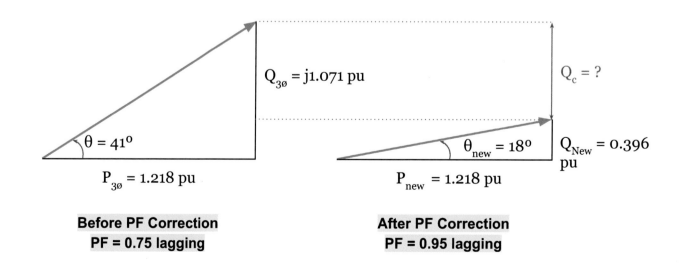

Before PF Correction **After PF Correction**
PF = 0.75 lagging PF = 0.95 lagging

46. The answer is: (B) 0.7

(← continued from previous page)

We can now calculate the amount of per unit reactive power that would be required to be supplied by a three-phase capacitor bank (Q_c) to correct the 480 volt bus to a new power factor of 0.95 lagging by taking the difference of the old reactive power ($Q_{3\phi}$) and the new reactive power (Q_{New}):

$$Q_C = Q_{3\phi} - Q_{New}$$
$$Q_C = j1.071 \ pu - j0.396 \ pu$$
$$Q_C = j0.675 \ pu$$

We can drop the imaginary j term since we know each of the possible answer choices are quantities of reactive power.

0.675 pu of three-phase reactive power would be required to be supplied to the 480 volt bus from a three-phase capacitor bank in order to improve the power factor of the bus from 0.75 lagging to 0.95 lagging.

The closest answer is 0.7.

The single line diagram of the power factor corrected system looks like this:

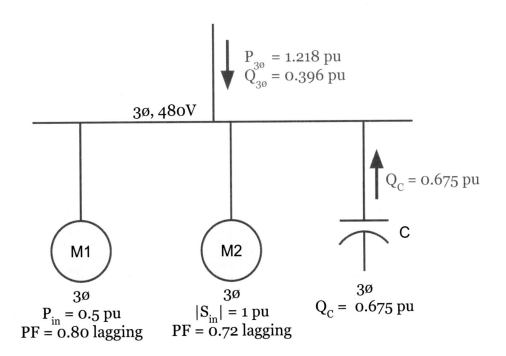

47. The answer is: (A) $143,000

A declining balance rate of 0.04 is a total period of 25 years:

$$d = \frac{1}{N}$$

$$0.04 = \frac{1}{N}$$

$$N = 25$$

Since the transformer depreciates by the same amount each year, we can assume straight line depreciation:

$$D = \frac{P-S}{N}$$

$$\$5,000 = \frac{P-\$18,000}{25}$$

$$P = \$143,000$$

The present worth value of the transformer is $143,000, the fact that the transformer was purchased 15 years ago is a red herring.

48. The answer is: (C) 8.5:5

First, let's calculate the three-phase transformer primary and secondary full load amps:

$$|I_{Pri\ FLA}| = \frac{5MVA}{\sqrt{3} \cdot 15kV} \qquad |I_{Sec\ FLA}| = \frac{5MVA}{\sqrt{3} \cdot 5kV}$$

$$|I_{Pri\ FLA}| = 192.5A \qquad |I_{Sec\ FLA}| = 577.4A$$

Next, let's calculate the current entering the ANSI #87 differential relay from the CTs located on the primary and secondary side of the three-phase transformer using the ratio of each set of CTs. Since CTs step down line current, multiply the line current by the CT ratio with the larger of the two numbers on the bottom of the fraction, and the smaller of the two numbers on the top of the fraction:

$$|I_{Pri\ CT}| = 192.5A\left(\frac{5}{200}\right) \qquad |I_{Sec\ CT}| = \sqrt{3}\,(577.4A)\left(\frac{5}{600}\right)$$

$$|I_{Pri\ CT}| = 4.8A \qquad |I_{Sec\ CT}| = 8.3A$$

Don't forget that the secondary CTs are delta connected. We will have to multiply the secondary CT current by an additional √3 multiplier *(see question #14 in this practice exam for a detailed explanation of the √3 multiplier for the delta connected CTs)*.

Prior to making any changes to the relay tap setting, the system looks like this:

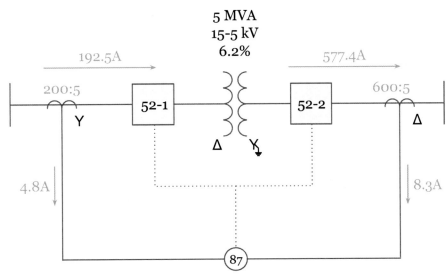

48. The answer is: (C) 8.5:5

(← continued from previous page)

Now let's calculate the optimal relay tap setting using the ratio of the two differential currents entering the ANSI #87 relay:

$$\frac{8.3A}{4.8A} = 1.7$$

We can now calculate the primary rating of the restraining winding by multiplying this ratio by 5 for a 5A secondary:

$$5(1.7) = 8.5$$

This means that the relay tap setting of the restraining winding in the relay when expressed as a CT ratio with a 5A secondary is 8.5:5.

We can verify this by stepping down the larger of the two relay currents using the restraining winding ratio:

$$8.3A\left(\frac{5}{8.5}\right) = 4.9A$$

Notice that after the larger of the two CT currents has been stepped down by the restraining winding, the two CT currents are almost identical (4.9A ≈ 4.8A). This is required to reduce differential mismatch as much as possible.

49. The answer is: (D) 640

First, let's determine the amount of illuminance measured in foot-candles in the room:

$$E = \frac{\Phi}{A}$$

$$E = \frac{1200 \; lm}{10ft \cdot 12ft}$$

$$E = 10fc$$

Now let's calculate the level of luminous intensity measured in candelas at the working plane 4 feet from the floor.

Remember that the pendant light is hung 2 feet from the ceiling so the distance from the light source to the working plane will equal the height of the room cavity:

$$E = \frac{I}{d^2}$$

$$I = (10fc)(14ft-2ft-4ft)^2$$
$$I = 640cd$$

There is a total of 640 candelas of luminous intensity at the working plane in the room.

50. The answer is: (C) 0.86

The average load is given, calculate the maximum connected load and plug it into the load factor formula:

$$L_{factor} = \frac{Average\ Load}{Maximum\ Load}$$

$$L_{factor} = \frac{20MVA}{115kVA + 225kVA + 10MVA + 3MVA + 10MVA}$$

$$L_{factor} = 0.86$$

The load factor for the new industrial customer is 0.86.

51. The answer is: (C) 476

To calculate torque, we will need to calculate the actual motor speed of the motor first. There are two poles in every pair of poles, P = 4 in the below formula for two pole pairs:

$$n_s = \frac{120f}{P}$$

$$n_s = \frac{120(60hz)}{4}$$

$$n_s = 1800rpm$$

$$n = n_s(1-s)$$

$$n = 1800rpm(1-0.08)$$

$$n = 1656rpm$$

The motor has a synchronous speed of 1800 rpm, and an actual speed of 1656 rpm. Let's use this information along with the rest of the given information to solve for torque:

Calculate the torque in lb·ft at rated conditions for a 60 Hz three-phase 150 HP motor that has two pole pair and a slip of 0.08.

$$T = \frac{33,000 \cdot Hp}{2\pi \cdot n}$$

$$T = \frac{33,000(150Hp)}{2\pi(1656rpm)}$$

$$T = 475.7 \ lb \cdot ft$$

52. The answer is: (D) All of the above.

Luminance is a measure of candela per area. Since the office is larger, the area is larger, resulting in a decrease in luminance. This quantity has changed.

The **lamp burnout factor** is a percentage usually determined by the lighting engineer to ensure that the lighting levels of the room are always met while allowing for a reasonable amount of lamps to be burned out. Since the new room is larger, it is likely that the lamp burnout factor has changed in order to meet the lighting requirements of the new room. This quantity has changed.

While the **coefficient of utilization** can account for specific losses associated with the lamp itself, it mostly accounts for losses due to the room the lamp is being utilized in such as the various surface reflectances present. Since the fixture has been moved to a larger room, this quantity has most likely changed.

53. The answer is: (A) The temporary receptacle must have ground fault circuit interrupter protection provided by a ground fault circuit interrupter receptacle, a ground fault circuit interrupter circuit breaker, or a ground fault circuit interrupter cord plugged directly into the temporary receptacle.

According to *NEC®590.6(A)*, a temporary circuit requires ground fault circuit interrupter protection for personnel.

There are three types of devices for ground fault circuit interrupter protection:

- A ground fault circuit interrupter (GFCI) protection receptacle.

    This type of receptacle has GFCI protection built into the receptacle and is commonly found in wet areas and bathrooms with a "push to test" button. A GFCI receptacle provides ground fault circuit interrupter protection for the receptacle itself, and all other receptacles wired to it.

    A picture of a GFCI receptacle is located in the *Handbook Edition of the 2017 National Electrical Code®* in article *210.8 Exhibit 210.8*.

- A ground fault circuit interrupter (GFCI) protection circuit breaker.

    This is a 120V panel circuit breaker with GFCI protection build into the circuit breaker. It is common to see non GFCI branch circuit receptacles that are fed from a GFCI panel circuit breaker to have a sticker that says "protected by GFCI" on the outlet to indicate that it is fed from a GFCI circuit breaker. A GFCI circuit breaker provides ground fault circuit interrupter protection for the entire circuit it feeds.

    A picture of a GFCI circuit breaker is located in the *Handbook Edition of the 2017 National Electrical Code®* in article *210.8 Exhibit 210.7*.

- A portable GFCI cord.

    GFCI cords can be plugged directly into non-GFCI receptacles to provide ground fault circuit interrupter protection. A GFCI cord provides ground fault circuit interrupter protection for any equipment that is plugged into the GFCI cord. *NEC®590.6(A)(1)* specifically mentions the permitted use of GFCI cords.

    A picture of a portable GFCI cord is located in the *Handbook Edition of the 2017 National Electrical Code®* in article *590.5 Exhibit 590.1*.

54. The answer is: (D) 1,500

First, let's start by drawing a simple one line diagram:

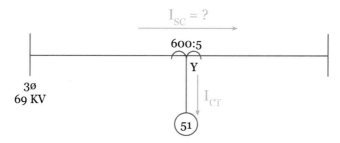

ANSI #51 time overcurrent relays have an inverse time curve on the TCC graph. This means that as the amperage value increases, the time delay to trip decreases, and as the amperage value decreases, the time delay to trip increases.

The greater the secondary CT current (I_{CT}) feeding the relay compared to the current tap setting (also known as the relay pick up or pick up current), the shorter the time delay is for the relay to operate. This comparison of the secondary CT current (I_{CT}) to the pick up current ($I_{Pick\ Up}$) is a ratio known as multiples of pick up (MPU), or multiples of current tap setting.

First, let's calculate the actual CT current (I_{CT}) at a pick up of 2.5 amps and 5 multiples of pick up:

$$MPU = \frac{I_{CT}}{I_{Pick\ Up}} \qquad I_{CT} = 2.5A(5)$$
$$I_{CT} = 12.5A$$

Next, let's use the CT ratio to step the secondary CT current (I_{CT}) value up to calculate the line current. Since the CT's are wye connected there is no additional √3 multiplier *(see sample exam question #14 solution for a detailed explanation of why)*. If a problem does not state or show the connection type (wye or delta) for a set of three CTs for a three-phase system, it is safe to assume they are wye connected.

Since we are stepping the current up from the secondary of the CT to the primary of the line, multiply the secondary CT current by the CT ratio with the larger of the two ratio numbers on the top of the fraction:

$$I_{SC} = 12.5A\left(\frac{600}{5}\right) = 1,500A$$

The short circuit current (fault current) on the line is 1,500 amps.

55. The answer is: (A) 17.2

First, let's draw out a single-phase equivalent circuit for a synchronous generator:

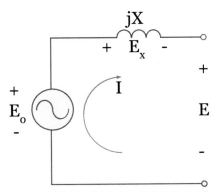

Synchronous Generator Single Phase Equivalent Circuit

Next, calculate the phase terminal voltage (E) from the three-phase voltage rating of the generator, and the line current at full load, and the power angle for a lagging 0.80 power factor:

$$|E| = \frac{13.8kV}{\sqrt{3}}$$

$$|E| = 7.967kV$$

$$|I| = \frac{55MVA}{\sqrt{3}\,(13.8kV)}$$

$$|I| = 2.301kA$$

$$\theta = \cos^{-1}(PF)$$
$$\theta = \cos^{-1}(0.80)$$
$$\theta = 37°$$

The last step before setting up a KVL equation to solve for the internal voltage (Eo) is to calculate the reactance of the generator in ohms by multiplying the per unit reactance by the base impedance:

$$Z_B = \frac{13.8kV^2}{55MVA}$$

$$Z_B = 3.463\Omega$$

$$X = X_{pu} \cdot Z_B$$
$$X = 1.4pu(3.463\Omega)$$
$$X = 4.848\Omega$$

We now have all of the information required to set up a KVL equation to solve for the internal voltage (Eo). Remember these are complex numbers, angles are important:

55. The answer is: (A) 17.2

(← continued from previous page)

$$\hat{E}_o = \hat{I} \cdot jX + \hat{E}$$

$$\hat{E}_o = (2.301kA\angle-37°)(j4.848\Omega) + 7.967kV\angle 0°$$

$$\hat{E}_o = 17.17kV\angle 31°$$

$$|E_o| = 17.17kV$$

The magnitude of the internal voltage is approximately 17.2kV.

Note that we were not given a voltage reference angle by the problem. We can set the power angle of the circuit (θ) equal to the line current (I) phase angle except opposite in polarity if we use a reference of zero degrees for the phase terminal voltage (E).

Don't forget to include the 90° phase angle (j) for the generator's reactance in your calculations.

56. The answer is: (B) 116

Typically, when sizing a motor disconnect, we use the tables in NEC® Article 430 based on motor voltage and horsepower for the full load current used in calculations and not the nameplate full load amps. However, when sizing a disconnect specifically for cranes, according to *NEC® 610.14(E)* the motor nameplate full load amps are used instead.

Generally there are two disconnects installed for cranes. The first is between the power supply and the runway conductors *NEC® 610.31,* and the second is between the runway conductors and the crane equipment *NEC® 610.32.*

For this question, we are tasked with sizing the disconnecting means of the runway conductors from the power supply. According to *NEC® 610.31,* the minimum rating of the disconnect will be not less *NEC® 610.14(E)(2)* for multiple motors.

For multiple motors, the motor load is calculated as the sum of the largest motor (or group of motors) followed by 50% of the sum of the next largest motor (or group of motors).

For a crane, a group of motors typically refers to motors that are switched on at the same time to move the crane in one direction, such as a pair of bridge motors that move the crane up and down the crane bridge.

Because the crane in this question has two bridge motors, the pair of bridge motors is counted as a single motor load of one group. 2 Bridge motors = 2 X 24A = 48A.

The largest motor load to be used for the calculation from *NEC® 610.14(E)(2)* is therefore the 92A hoist motor, followed by the 48A group of bridge motors as the next largest motor load. The 31A trolley motor will not be used in the calculation since it is smaller than the second largest motor (or group of motors).

The minimum ampacity rating of the runway conductor disconnect can be calculated as:

Answer: 92A + 50%(48A) = 116A.

Note that the problem asked for the minimum ampacity rating of the disconnect, and not the actual standard size of the disconnect. This means we are just calculating what the minimum ampacity rating of the disconnect needs to be and not rounding up to meet a standard disconnect rating.

57. The answer is: (A) -2.0

Don't be afraid of working problems in the per unit system. Notice that if this problem was given in actual values instead of per unit, it would appear that we are actually missing information to calculate the voltage regulation, such as the generator rated values.

However, since we are working in the per unit system with base values equal to the machine's ratings, we don't actually need to know the volt-amp or voltage ratings.

Since the base voltage is equal to the machine's voltage output rating, the magnitude of the single-phase terminal voltage (E) is equal to 1 per unit. Since the base current is equal to the machine's full load amps, the magnitude of the current supplied by the machine at full load (I) is equal to 1 per unit. In both cases, this is because the base value is equal to the actual value (Vb/V = 1 pu and Ib/I = 1 pu).

Let's start by calculating the power angle (θ) at a leading power factor of 0.83. Don't forget to make theta negative since it is leading:

$$\theta = \cos^{-1}(0.83) = -34°$$

We can change the polarity of the power angle and assign it to the phase current angle If we use a reference of zero degrees for the single-phase terminal voltage (E) *(see note 1)*.

We now have enough information to drawn and fill in a single-phase equivalent circuit for the three-phase generator. Don't forget to include a "j" for the reactance given by the problem:

Let's calculate what the internal (or "induced") phase voltage of the generator (E_o) would have to be in order for the machine to still be able to output the rated terminal voltage (E = 1pu<0°) when the generator is operating at full load and a 0.83 leading power factor:

57. The answer is: (A) -2.0

(← continued from previous page)

$$\hat{E}_o = \hat{E}_x + \hat{E}$$
$$\hat{E}_o = (1pu\angle 34°)(j0.04pu) + 1pu\angle 0°$$
$$\hat{E}_o = 0.978pu\angle 2°$$

We are now ready to calculate the voltage regulation of the machine at full load and a leading power factor of 0.83. Even though we had to properly include all complex phase angles above to calculate the internal phase voltage of the generator (Eo), voltage regulation is a magnitude comparison only.

This means that we will drop the angle for both the internal (Eo) and terminal voltage (E) and use only magnitudes:

$$V_{Reg\ \%} = \frac{|E_o|-|E|}{|E|} \cdot 100$$

$$V_{Reg\ \%} = \frac{0.978pu - 1pu}{1pu} \cdot 100$$

$$V_{Reg\ \%} = -2.2\%$$

The voltage regulation of the machine is -2.2%. The closest answer is -2.0 percent.

Note that a leading synchronous generator typically has a negative voltage regulation while a lagging synchronous generator typically has a positive voltage regulation.

Note 1: Refer to the Leading and Lagging Cheat Sheet in the Free Articles section of www.electricalpereview.com for a more in depth explanation on changing the polarity of the power angle and assigning it to the phase current angle by using a reference of zero degrees for the phase voltage angle:

Electrical PE Review - Leading Lagging Cheat Sheet:
https://www.electricalpereview.com/leading-lagging-cheat-sheet/

58. The answer is: (C) 96%

The efficiency (η) of an electric machine is defined as the ratio of the output power (P_{out}) delivered to the load, to the input power (P_{in}) drawn from the power supply.

The sum of the total power losses of a machine (P_{loss}) and the output power (P_{out}) will equal the input power (P_{in}) drawn from the power supply and can be substituted into the efficiency (η) formula:

$$\eta = \frac{P_{out}}{P_{in}}$$

$$\eta = \frac{P_{out}}{P_{out} + P_{loss}}$$

We can determine the output power (P_{out}) of the transformer at rated conditions using the given apparent power |S| rating of the transformer and power factor (PF):

$$P_{out} = |S| \cdot PF$$
$$P_{out} = (75kVA)(0.82)$$
$$P_{out} = 61.5kW$$

We will need to determine the total losses of the machine from the results of the open circuit and short circuit test. There are two types of losses in a transformer.

The first type of losses are the copper losses (P_{cu}). Copper losses are sometimes referred to as "heat" losses, "winding" losses, or "I²R" losses instead. Copper losses can be determined from the short circuit test.

The second type of losses are the core losses (P_c). Core losses are sometimes referred to as "no load" losses, or "magnetizing" losses instead. Core losses can be determined from the open circuit test.

The sum of both the copper losses (P_{cu}) and core losses (P_c) will equal the total losses of the transformer (P_{loss}):

$$P_{loss} = P_{cu} + P_c$$

58. The answer is: (C) 96%

(← continued from previous page)

The current drawn by the **short circuit test** (I_{sc}) is almost entirely from the series winding resistance (R) and winding reactance (X), allowing us to effectively neglect the parallel magnetizing core branch:

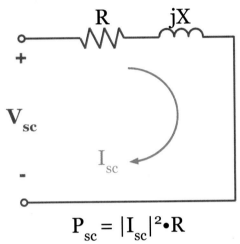

$$P_{sc} = |I_{sc}|^2 \cdot R$$

Transformer Short Circuit Test

Copper losses (P_{cu}) vary with load, but since the amps measured on the secondary side of the transformer during the short circuit test (I_{sc}) were equal to the secondary rated current of the transformer, the power measured during the short circuit test (P_{sc}) will equal 100% of the copper losses (P_{cu}), or, $P_{sc} = P_{cu} = 1,900$ W.

The current drawn by the **open circuit test** (I_{oc}) is almost entirely from the parallel magnetizing resistance (R_m) and magnetizing reactance (X_m), this time allowing us to effectively neglect the series winding resistance and reactance:

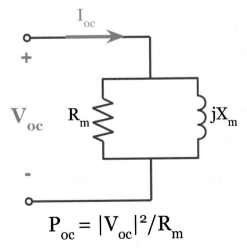

$$P_{oc} = |V_{oc}|^2 / R_m$$

Transformer Open Circuit Test

58. The answer is: (C) 96%

(← continued from previous page)

Unlike copper losses (P_{cu}), core losses (P_c) do not vary with load, and since rated voltage was applied to the transformer during the open circuit test (V_{oc}) we know that the power measured during the open circuit test (P_{oc}) will equal the transformer core losses (P_c) at rated conditions, or: $P_{oc} = P_c = 380$ W.

We can now add the transformer copper losses (P_{cu}) and the transformer core losses (P_c) to calculate the total transformer losses (P_{loss}):

$$P_{loss} = P_{cu} + P_c$$
$$P_{loss} = 1.9kW + 380W$$
$$P_{loss} = 2.28kW$$

Now that we know both the output power of the transformer at full load and 0.82 power factor, and the losses at full load, we can calculate the efficiency of the transformer:

$$\eta = \frac{61.5kW}{61.5kW + 2.28kW}$$

$$\eta = 0.9643$$
$$\eta = 96.4\%$$

The efficiency of the transformer is approximately 96.4%.

Electrical Power PE Practice Exam
9. Afternoon Session Solutions

59. The answer is: (A) The real power and voltage magnitude of the bus are inputs, the reactive power and voltage angle of the bus in reference to the slack bus are calculated by the program.

Each bus in a system can be classified into three different types when analyzing the power flow:

1. Slack bus - sometimes referred to as a "swing bus".
2. Voltage controlled bus - sometimes referred to as a "PV bus".
3. Load bus - sometimes referred to as a "PQ bus".

1. Slack bus (swing bus) - the voltage controlled bus that is designed as the reference bus for the rest of the system. There is only one slack bus for the entire system being modeled. Typically the most upstream power source or generator bus is chosen as the slack bus.

The input data for the slack bus is the voltage magnitude of the bus (V) and the voltage phase angle of the bus (δ). Typically, a reference of zero degrees is chosen for the voltage phase angle of the slack bus to make it easier to interpret the voltage phase angles of all load and voltage controlled busses in the system.

The software modeling program calculates the real power of the slack bus (P) and the reactive power of the slack bus (Q).

2. Voltage controlled bus (PV bus) - any bus that generates either real power (P), reactive power (Q), or both such as a bus connected to a generator or capacitor bank.

The input data for a voltage controlled bus (PV bus) is the real power of the bus (P), and the voltage magnitude of the bus (V).

The software modeling program calculates the reactive power of the voltage controlled bus (Q), and voltage phase angle of the voltage controlled bus (δ) in reference to the voltage phase angle of the slack reference bus.

3. Load bus (PQ bus) - any bus that consumes either real power (P), reactive power (Q), or both such as a bus that delivers power to a motor or to a lighting panel.

The input data for a load bus (PQ bus) is the real power of the bus (P), and the reactive power of the bus (Q).

The software modeling program calculates the voltage magnitude of the load bus (V), and voltage phase angle of the load bus (δ) in reference to the voltage phase angle of the slack reference bus.

60. The answer is: (D) 1.4+j0.2Ω

We are not told if the unbalanced load is delta or wye connected, however, we can still calculate the total three-phase power drawn by the unbalanced load by summing together the power drawn in each individual unbalanced phase.

First, calculate the power angle of each phase:

Phase A power angle:

$\theta = \cos^{-1}(0.97)$

$\theta = 14°$

Phase B power angle:

$\theta = \cos^{-1}(0.81)$

$\theta = 36°$

Phase C power angle:

$\theta = \cos^{-1}(0.72)$

$\theta = -44°$

(The power angle for phase C is negative since it has a leading power factor).

Next, let's calculate the total complex three-phase power consumed by the unbalanced load by summing together the complex power consumed in each unbalanced phase:

$$\hat{S}_{3\phi} = \hat{S}_{A\phi} + \hat{S}_{B\phi} + \hat{S}_{C\phi}$$

$$\hat{S}_{3\phi} = 75kVA\angle 14° + 65kVA\angle 36° + 45kVA\angle -44°$$

$$\hat{S}_{3\phi} = 159.7kVA\angle 9°$$

Last, let's calculate the equivalent impedance of a balanced and wye connected load that consumes the same amount of power from a 480V three-phase power source. We can calculate the magnitude of the impedance first, and then take the impedance angle directly from the power angle to form the complex impedance:

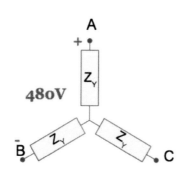

$|S_{3\phi}| = \dfrac{|V_L|^2}{|Z|}$

$|Z| = \dfrac{(480V)^2}{159.7kVA}$

$|Z| = 1.4\Omega$

$\hat{Z} = 1.4\Omega\angle 9°$

$\hat{Z} = 1.38 + j0.23\Omega$

60. The answer is: (D) 1.4 + j0.2Ω

(← continued from previous page)

Alternatively, we could have solved for the same balanced wye impedance using phase values instead:

$$|S_{1\phi}| = \frac{|V_p|^2}{|Z|} \qquad |Z| = 1.4\Omega$$

$$|Z| = \frac{(\frac{480V}{\sqrt{3}})^2}{\frac{159.7kVA}{3}} \qquad \hat{Z} = 1.4\Omega < 9°$$

$$\hat{Z} = 1.38 + j0.23\Omega$$

Lastly, we could also have solved for the impedance using a more roundabout way using Ohm's law and complex power by first determining the current drawn in each phase by the equivalent balanced wye connected load using the single phase complex power formula:

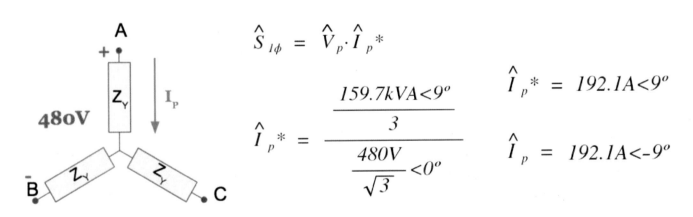

$$\hat{S}_{1\phi} = \hat{V}_p \cdot \hat{I}_p{}^*$$

$$\hat{I}_p{}^* = \frac{\frac{159.7kVA<9°}{3}}{\frac{480V}{\sqrt{3}}<0°}$$

$$\hat{I}_p{}^* = 192.1A<9°$$

$$\hat{I}_p = 192.1A<-9°$$

Or by using the three-phase apparent power formula, the wye relationship that $I_p = I_L$ and setting the current angle equal to the power angle except opposite in polarity:

$$|S_{3\phi}| = \sqrt{3} \cdot |V_L||I_L| \qquad |I_L| = 192.1A$$

$$|I_L| = \frac{159.7kVA}{\sqrt{3}(480V)} \qquad \hat{I}_L = 192.1A<-9°$$

60. The answer is: (D) 1.4 + j0.2Ω

(← continued from previous page)

Now that the current drawn by the equivalent balanced wye load is known, we can use Ohm's law to solve for impedance:

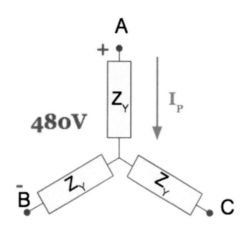

$$\hat{V}_p = \hat{I}_p \cdot \hat{Z}_Y$$

$$\hat{Z}_Y = \frac{\frac{480V}{\sqrt{3}} < 0°}{192.1A < -9°}$$

$$\hat{Z}_Y = 1.42 + j0.23\,\Omega$$

The impedance of a balanced and wye connected load that draws the same amount of three-phase power as unbalanced load is 1.4 + j0.2Ω.

61. The answer is: (D) 113

Magnetizing reactance (X_m) is the reactance in the parallel magnetizing branch and leakage reactance (x) is the reactance in the series stator impedance branch:

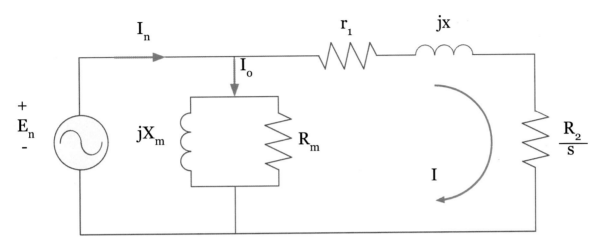

Induction Motor Single Phase Equivalent Circuit

During the locked rotor test, the magnetizing branch is negligible because the current flowing through the series stator and rotor circuit (I) is much greater than the current flowing through the magnetizing branch (I_o):

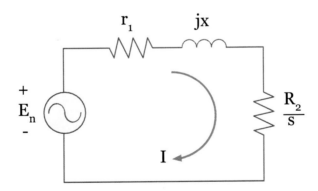

Induction Motor Single Phase Equivalent Circuit Locked Rotor with Magnetizing Branch Removed

The real power is already given to us by the problem statement. To calculate the total apparent power, we'll need to calculate the reactive power first. We will use the leakage reactance (x) only in our calculations since only reactive impedance absorbs reactive power (the other two resistors would absorb real power).

Since it is a series circuit, the current flowing through each impedance is the same. Because of this, we can use the I^2Z equation for power:

61. The answer is: (D) 113

(← continued from previous page)

$$Q_{3\phi} = 3 \cdot Q_{1\phi}$$
$$Q_{3\phi} = 3 \cdot |I|^2 X$$
$$Q_{3\phi} = 3 \cdot (200A)^2 (0.49\Omega)$$
$$Q_{3\phi} = 58.8 kVAR$$

Now let's use the Pythagorean theorem along with the given real power to solve for the apparent power:

$$|S_{3\phi}| = \sqrt{P_{3\phi}^2 + Q_{3\phi}^2}$$
$$|S_{3\phi}| = \sqrt{96.5 kW^2 + 58.8 kVAR^2}$$
$$|S_{3\phi}| = \sqrt{96.5 kW^2 + 58.8 kVAR^2}$$
$$|S_{3\phi}| = 113.0 kVA$$

Alternatively, we could have used our calculator to convert the complex power in rectangular form to polar form to:

$$\hat{S}_{3\phi} = P_{3\phi} + jQ_{3\phi}$$
$$\hat{S}_{3\phi} = 96.5 kW + j58.8 kVAR$$
$$\hat{S}_{3\phi} = 113.0 kVA < 31°$$
$$|S_{3\phi}| = 113.0 kVA$$

The total apparent power drawn by the motor during locked rotor conditions is 113kVA.

9. Afternoon Session Solutions

62. The answer is: (B) 2 $\frac{1}{2}$

For simpler raceway sizing problems when the type and size of the conductors are the same, we can jump straight to the tables in *NEC® Annex C* to determine the minimum size raceway based on raceway type.

However, when the conductors in the raceway have different sizes and/or insulation types, the total individual cross section area of each conductor in the raceway must first be calculated using the Tables in *NEC® Chapter 9*.

Step 1: Determine the cross section area of each conductor in the raceway. For RHH without the outer covering appears in *NEC® Ch. 9* Table 5 as RHH* (RHH without the asterisk is with the outer covering on):

RHH* 1/0 AWG = 0.2223 in^2
RHH* 4/0 AWG = 0.3718 in^2

According to the problem, there are **three** 4/0 AWG RHH* conductors, and **one** 1/0 AWG RHH* conductor in the raceway. Calculate the total conductor area:

Total conductor cross section area in raceway = 3(0.3718 in^2) + 0.2223 in^2
Total conductor cross section area in raceway = 1.3377 in^2

Step 2: Check with *NEC® Ch. 9 Table 1* to determine the how much bigger the raceway needs to be compared to the *total conductor cross section area* according to how many conductors are in the raceway:

For more than 2 conductors in the raceway, the raceway must be **40%** larger than the *total conductor cross section area*.

Step 3: Find the correct table in *NEC® Ch. 9 Table 4* for the raceway type. The raceway type in the problem is liquidtight flexible metal conduit, so we want to use the table titled: *Article 350 — Liquidtight Flexible Metal Conduit (LFMC)*.

Using the Over 2 Wires 40% column in inches squared (in.^{2}), round up to the next area size compared to the *total conductor cross section area* of 1.3377 *in^2*.

We are rounding up instead of down since this represents the minimum size diameter raceway, as in the raceway cannot be any smaller than this number, and the calculated number does not match a standard size.

The next size up is 1.953 in^2 which corresponds to the maximum 40% total conductor area for size 2 $\frac{1}{2}$ in. LFMC raceway. Continued on next page →

62. The answer is: (B) 2 $1/2$

(← continued from previous page)

Step 3 Alternative: Another way to solve this is to calculate the actual minimum size area of the raceway based off the *total conductor cross section area* and the multiplier from *NEC® Ch. 9 Table 1*:

Total conductor cross section area in raceway from step 1 = 1.3377 in^2

NEC® Ch. 9 Table 1 = raceway must be **40%** larger than the *total conductor cross section area* from step 2.

Minimum cross section area of the raceway = 1.3377 in^2 / 40%
Minimum cross section area of the raceway = 3.3443 in^2

Now go back to *NEC® Ch. 9 Table 4* Article 350 — Liquidtight Flexible Metal Conduit (LFMC).

This time use the Total Area 100% column in inches squared (in.^{2}) and round up to the next area size compared to the *minimum cross section area of the raceway* 3.3443 in^2.

The next size up is 4.881 in^2 which corresponds to the actual diameter size of the 2 $1/2$ inch LFMC raceway.

63. The answer is: (D) 300:5

The most common industry standard for CT ratios is a maximum of 5 amps on the secondary of the CT.

Since each answer choice reflects this, we will have to include the 5 Amps secondary in our calculation.

Solve for the primary max current of the CT if the secondary max current is equal to 5 amps:

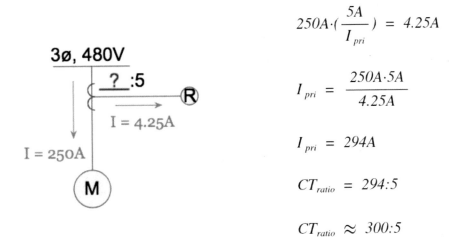

$$250A \cdot \left(\frac{5A}{I_{pri}}\right) = 4.25A$$

$$I_{pri} = \frac{250A \cdot 5A}{4.25A}$$

$$I_{pri} = 294A$$

$$CT_{ratio} = 294:5$$

$$CT_{ratio} \approx 300:5$$

The closest answer is a CT ratio of 300:5.

64. The answer is: (A) Capacitive Charging Current.

Capacitive charging current (I_c) is due to the capacitive nature of insulation that becomes charged in the same manner as the dielectric medium of a capacitor:

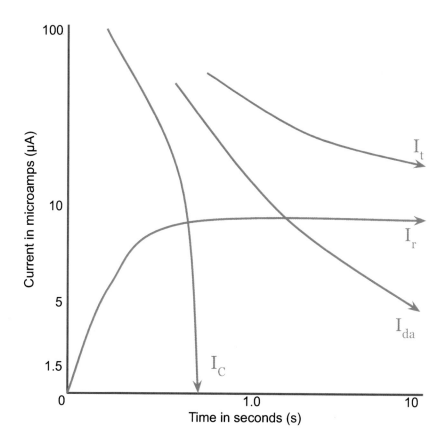

Capacitive Charging Current (I_c) - The charging current drawn by the capacitive properties of electrical insulation.

Dielectric Absorption Current (I_{da}) - The current drawn by the polarization properties of electrical insulation material.

Leakage Conduction Current (I_r) - The current drawn through and over electrical insulation due to the resistive properties of the material.

Total Insulation Current (I_t) - The sum of all three different currents drawn by the material of the insulation itself.

As voltage is applied, the **capacitive charging current (I_c)** reaches its maximum value almost immediately, then drops off, or decays rather quickly as the capacitive impedance of the insulation becomes fully charged and no longer able to draw current. Note that there is no such thing as *Absorption Conduction Current* (answer choice D), this was a trick answer.

65. The answer is: (A) 3 ½

Working space clearances for 600 volts or less is given in *NESC® Table 125-1*.

Exposed energized parts on one side and grounded parts on the other side is classified as *Condition 2*.

277 Volts to ground falls in the *151 - 600 volts to ground* category.

The table value for condition 2 clearance and 151 - 600 volts to ground is 3 ½ feet.

66. The answer is: (D) 11 + j5

First, draw the three-phase wye load:

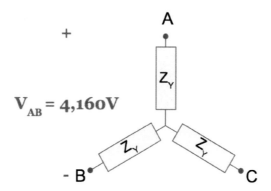

Next, let's calculate the power angle from the given power factor. The power angle is positive since the power factor is lagging:

$$\theta = \cos^{-1}(0.92)$$
$$\theta = 23°$$

Now let's combine the power angle and the given three-phase apparent power drawn by the load to determine the three-phase complex power drawn by the load:

$$\hat{S}_{3\theta} = 1{,}500kVA \angle 23°$$

To calculate the wye connected impedance, we can use the line voltage and apparent power setting the impedance angle equal to the power angle:

$$|Z_Y| = \frac{|V_L^2|}{|S_{3\phi}|}$$

$$|Z_Y| = \frac{(4{,}160V)^2}{1{,}500kVA}$$

$$|Z_Y| = 11.5\Omega$$

$$\hat{Z}_Y = 11.5\Omega \angle 23°$$

$$\hat{Z}_Y = 10.6 + j4.5\Omega$$

66. The answer is: (D) 11 + j5

(← continued from previous page)

Alternatively, we can solve for the load impedance using ohm's law. To do this, we will need to determine the phase current drawn by each phase of the wye connected load:

$$|S_{1\phi}| = |V_p| \cdot |I_p|$$

$$|I_p| = \frac{|S_{1\phi}|}{|V_p|}$$

$$|I_p| = \frac{\left(\frac{1,500kVA}{3}\right)}{\left(\frac{4,160V}{\sqrt{3}}\right)}$$

$$|I_p| = 208.2A$$

With the phase current drawn by the load and the phase voltage across the load known, we can calculate the load impedance using the power angle for the load impedance:

$$|V_p| = |I_p| \cdot |Z_Y|$$

$$|Z_Y| = \frac{\left(\frac{4,160V}{\sqrt{3}}\right)}{208.2A}$$

$$|Z_Y| = 11.5\Omega$$

$$\hat{Z}_Y = 11.5\Omega \angle 23°$$

$$\hat{Z}_Y = 10.6 + j4.5\Omega$$

The impedance of the wye connected load is 10.6+j4.5Ω.

67. The answer is: (B) Heat losses will be approximately 150% compared to full load.

Heat losses (P_{cu}) are related to the square of the change in percent load. We can calculate the heat losses, also known as copper or I^2R losses using the following formula:

$$P_{cu} = I^2R$$

When the transformer is operating at rated nameplate conditions, it is operating at full load, or 100% loading conditions. The heat losses at full load are equal to I^2R:

$$P_{cu\ 100\%\ Load} = (100\%I)^2R$$
$$P_{cu\ 100\%\ Load} = 100\%^2 \cdot I^2R$$
$$P_{cu\ 100\%\ Load} = I^2R$$

At 25% above full load, or 125% loading conditions, the heat losses increase to 156.25% compared to what they were at full load:

$$P_{cu\ 125\%\ Load} = (125\%I)^2R$$
$$P_{cu\ 125\%\ Load} = 125\%^2 \cdot I^2R$$
$$P_{cu\ 125\%\ Load} = 1.5625 \cdot I^2R$$
$$P_{cu\ 125\%\ Load} = 156.25\% \cdot I^2R$$

The closest answer is 150%.

68. The answer is: (A) 300

Regardless of what the relay input current is that the relay is set to trip for (known as the tap setting, or pick up current), the CT ratio is generally selected so that the maximum load of the circuit comes close to, but does not exceed the 5 amps secondary rating of the CT when it is stepped down. This means that the 250% primary trip is a red herring.

Let's draw a diagram of the system:

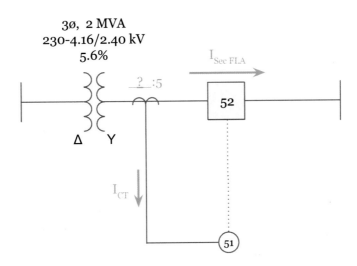

Since we are dealing with secondary protection, let's calculate the secondary full load amps that the transformer will draw from the circuit:

$$|I_{Sec\ FLA}| = \frac{|S_{3\phi}|}{\sqrt{3}|V_L|} = \frac{2MVA}{\sqrt{3}(4.16kV)} = 277.6A$$

The secondary full load amps drawn by the transformer is 277.6 amps. A CT ratio of 277.6:5 would result in 5 amps secondary. However, since 277.6:5 is not a standard CT ratio, nor is it a possible answer choice, we must choose the next size up of 300:5. At full load, this would result in a relay input current of 4.6 amps:

$$I_{CT} = 277.6A\left(\frac{5}{300}\right) = 4.6A$$

The correct rating for the primary of the CT out of the choices given for a CT with a 5 amp rated secondary is 300 amps. This would result in a CT ratio of 300:5.

69. The answer is: (B) 13.8

Draw the circuit, fill in the given values, and write down the relationships for a single line to ground fault:

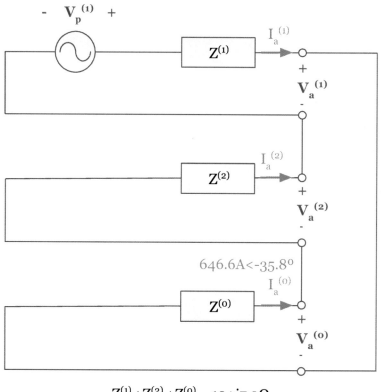

Symmetrical Components SLG Fault Single Phase Equivalent Circuit

For a single line to ground fault, the sequence component currents are equal to each other: $I_a^{(1)} = I_a^{(2)} = I_a^{(0)}$. Let's use this information to solve for the system line to neutral voltage $V_p^{(1)}$:

$$\hat{V}_p^{(1)} = (\hat{I}_a^{(0)})(\hat{Z}_{eq})$$

$$\hat{V}_p^{(1)} = (646.6A\angle{-35.8°})(10+j7.2\Omega)$$

$$\hat{V}_p^{(1)} = 7.97kV \angle 0°$$

$$|V_L| = \sqrt{3} \cdot |V_p^{(1)}|$$
$$|V_L| = \sqrt{3} \cdot (7.97kV)$$
$$|V_L| = 13.8kV$$

Since the question asks for the magnitude of the line voltage, we will need to multiply the magnitude of system line to neutral voltage $V_p^{(1)}$ by √3. The line voltage of the system is 13.8kV.

70. The answer is: (A) The synchronous reactance voltage drop leads the stator current by 90 degrees for both a synchronous motor and synchronous generator.

We can determine this by evaluating the voltage drop across the reactance of each machine using the single-phase equivalent circuit for both a synchronous motor and synchronous generator:

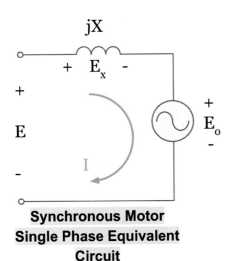

Synchronous Motor Single Phase Equivalent Circuit

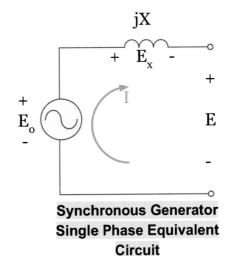

Synchronous Generator Single Phase Equivalent Circuit

Synchronous motor voltage drop (E_x):

$$\hat{E}_x = \hat{I} \cdot jX$$
$$\hat{E}_x = (|I|<\theta_I)(X<90°)$$
$$\hat{E}_x = (|I| \cdot X)<(\theta_I+90°)$$

$$\theta_{Ex} = \theta_I + 90°$$

Synchronous generator voltage drop (E_x):

$$\hat{E}_x = \hat{I} \cdot jX$$
$$\hat{E}_x = (|I|<\theta_I)(X<90°)$$
$$\hat{E}_x = (|I| \cdot X)<(\theta_I+90°)$$

$$\theta_{Ex} = \theta_I + 90°$$

Regardless of what the stator current phase angle (θ_I) is, the voltage drop (E_x) across the synchronous reactance will always lead the stator current phase angle by 90 degrees for both a synchronous motor and synchronous generator.

(note how the voltage drop polarity of E_x is the same with respect to the direction of the stator current for both a synchronous motor and synchronous generator in the circuits shown above).

71. The answer is: (A) Increase the long time tap setting for CB-A.

First, let's define each of the three relay trip functions on the TCC graph:

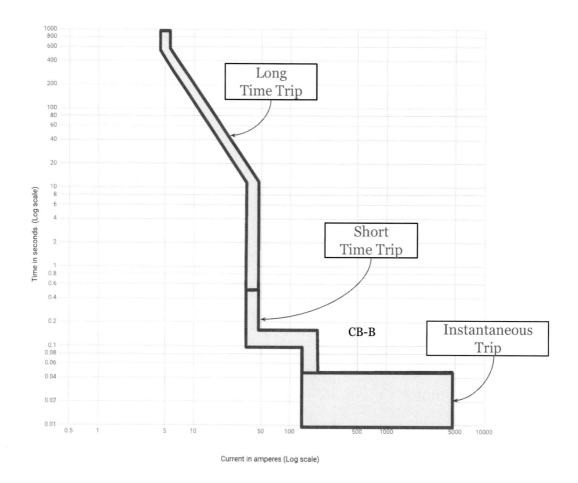

The **instantaneous** trip function generally occurs from 0.05 seconds and less, the **short time** trip function generally occurs between 0.05 and 0.5 seconds, and the **long time** trip function generally occurs between 0.5 and 1,000 seconds. The above TCC graph is in the logarithmic scale for both the x and y axis.

A change in the **horizontal current axis** will increase or decrease the tap setting, also known as the pick up current. This is the amperage that the relay begins to initiate operation at.

A change in the **vertical time axis** will increase or decrease the time delay setting, also known as the time dial setting. This is the duration of time that the current must sustain before the relay begins to initiate operation.

For example, let's look at the four possible changes we could make to CB-B's long time trip function:

71. The answer is: (A) Increase the long time tap setting for CB-A.

(← continued from previous page)

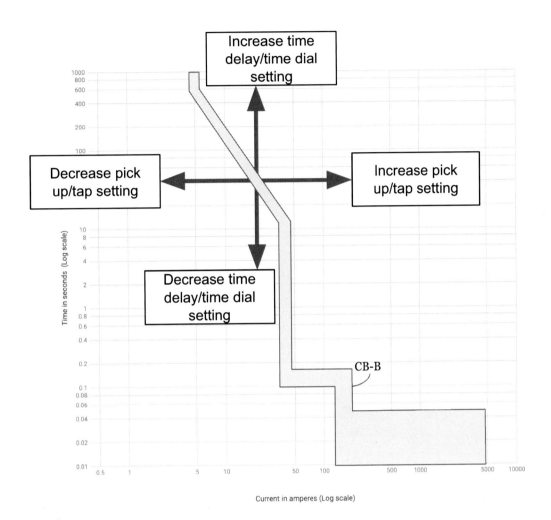

In the TCC graph in the problem, the system is not coordinated properly due to the overlap between CB-A and CB-B's long time trip region.

To properly coordinate the system, we need to adjust either CB-A or CB-B's **time current curve (TCC)** so that there is no overlap.

We must also make sure that CB-A's time current curve is completely above and to the right of CB-B's time current curve since CB-A is upstream of CB-B.

Let's evaluate each possible answer choice to determine the correct answer:

71. The answer is: (A) Increase the long time tap setting for CB-A.

(← continued from previous page)

(A) Increase the long time tap setting for CB-A.

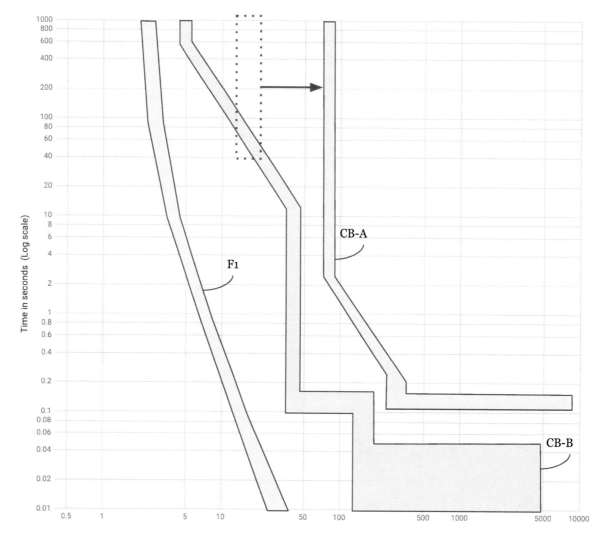

Increasing the long time tap setting for CB-A could potentially meet our requirements as long as it is increased far enough to completely clear CB-B's long time trip region.

Compared to the other following possible answers, this is the only change in device settings that would properly coordinate the system. **This is the correct answer.**

71. The answer is: (A) Increase the long time tap setting for CB-A.

(← continued from previous page)

(B)   Enabling instantaneous trip for CB-A.

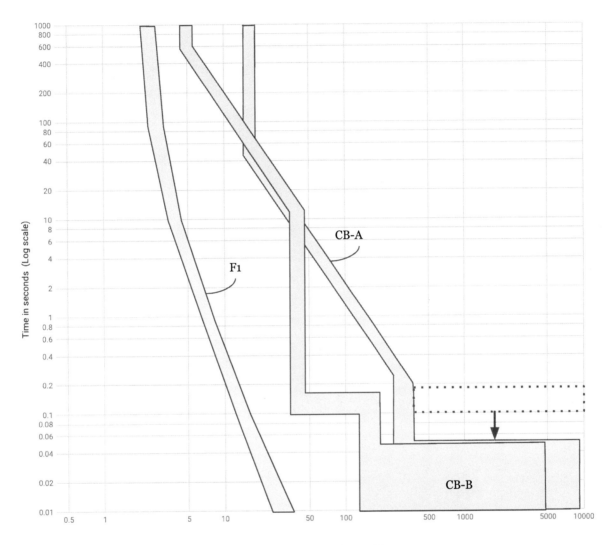

Enabling, or turning on the instantaneous trip function for CB-A does not remove the overlap with CB-B's long time trip function and may potentially increase the level of miscoordination by introducing an additional overlap between CB-A's and CB-B's instantaneous trip functions as well.

This is not the correct answer.

71. The answer is: (A) Increase the long time tap setting for CB-A.

(← continued from previous page)

(C)   Increase the short time delay setting for CB-B.

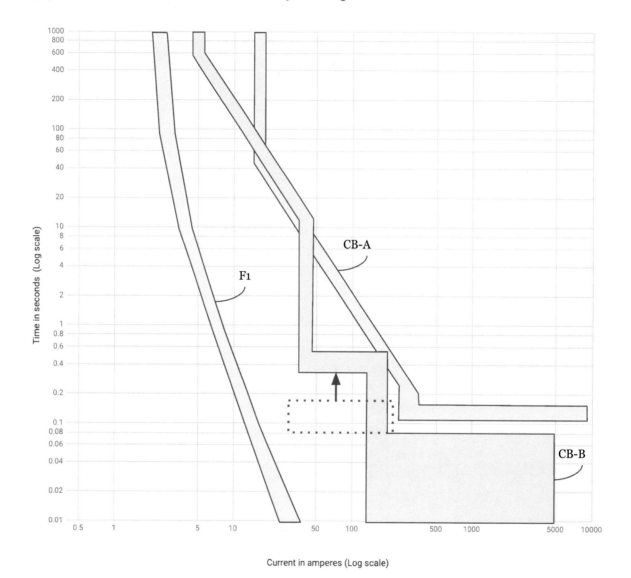

Increasing the short time delay for CB-B does not remove the overlap with CB-A's long time trip function, and may potentially increase the level of miscoordination by introducing an additional overlap between CB-B's and CB-A's short time trip functions as well.

This is not the correct answer.

71. The answer is: (A) Increase the long time tap setting for CB-A.

(← continued from previous page)

(D) Decrease the long time tap setting for CB-A.

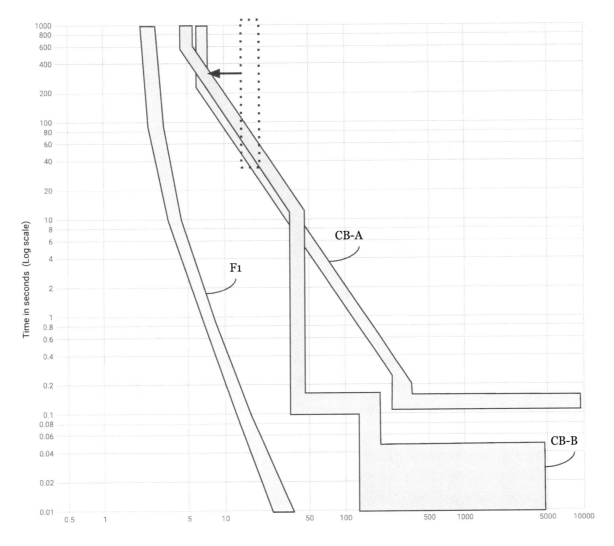

Decreasing the long time tap setting for CB-A only further increases the level of overlap with CB-B's long time trip function.

This is not the correct answer.

72. The answer is: (C) Series compensation may be used to improve transmission power transfer.

The word compensation generally refers to the addition of reactive devices to a transmission line in order to balance the impedance of the system when it is too inductive or capacitive to improve system stability, voltage drop, power factor, or power transfer.

Compensators can be made up of inductive reactance, capacitive reactance, or an adjustable combination of both such as a static var compensator.

Shunt (parallel) compensators are mostly used to help supply reactive power to an inductive circuit.

Series (in line) compensators are mostly used to help balance the overall impedance of the transmission line conductors so that it is less inductive.

Transmission line power transfer:

$$P = \frac{|V_S| \cdot |V_R|}{X} sin(\delta)$$

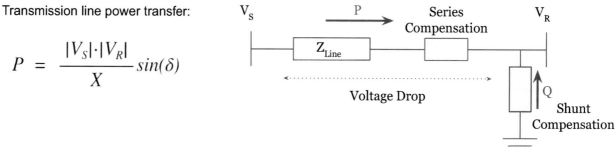

The typical advantages of compensation are:

Improve (increase) the available voltage magnitude at the receiving end (V_R) due to less (decreased) voltage dropped along the line.

Improve the system stability by decreasing the angle difference (δ) between the sending (V_S) and receiving (V_R) voltage.

Improve (increase) the available transmission power transfer (P) by decreasing the angle difference between the sending and receiving voltage (δ), and by decreasing the overall line reactance (X_L in $Z_{Line} = R + jX_L$).

Improve the power factor of the transmission line by reducing the amount of reactive power required to be supplied by the transmission line due to supplemental reactive power (Q) supplied by shunt compensation.

Compensators are typically not used as an application to negate the effects of harmonics. Such devices are referred to as filters or conditioners.

73. The answer is: (C) 230

The volts per hertz ratio is responsible for the amount of flux in a motor.

First let's calculate the volts per hertz ratio that exists at rated conditions:

$$\frac{460V}{60Hz} = 7.67 \ V/Hz$$

Now let's use this information to calculate what voltage is required to maintain the same V/Hz ratio at a frequency of 50%

$$\frac{V}{f} = 7.67 \ V/Hz$$

$$V = (7.67 \ V/Hz)(f)$$

$$V = (7.67 \ V/Hz)(50\% \cdot 60Hz)$$

$$V = 230 \ Volts$$

230 Volts are required to maintain the same 7.67 V/Hz flux ratio.

74. The answer is: (B) Transmission line 2 protection

If a fault occurs within the **utility transmission bus protection zone**, the main breaker and feeder breakers connected to the utility transmission bus will operate in order to isolate the bus. The utility transmission bus and every circuit downstream of it will be de-energized as a result:

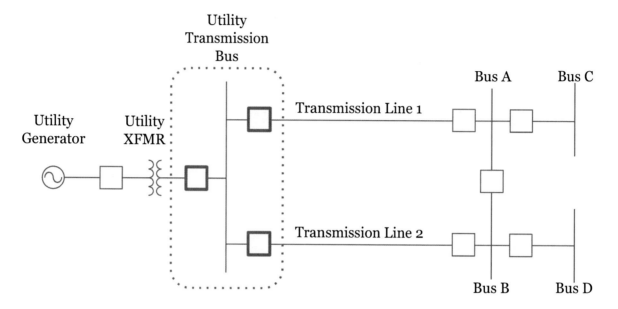

If a fault occurs within the **transmission line 2 protection zone,** both the sending and receiving breakers will operate in order to isolate the line. Transmission line 2 will be de-energized, however bus D remains energized due to the tie breaker connecting bus A and bus B **(this is the correct answer)**:

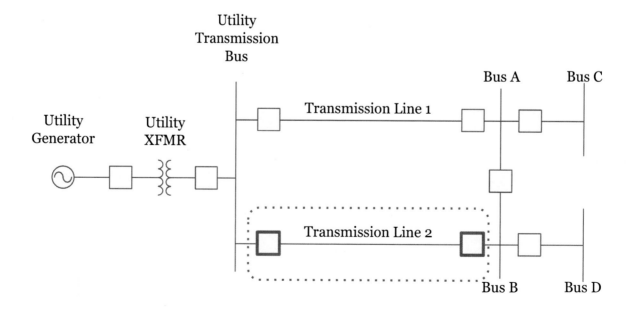

74. The answer is: (B) Transmission line 2 protection

(← continued from previous page)

If a fault occurs within the **utility transformer differential protection zone**, the primary and secondary transformer breakers will operate in order to isolate the transformer. The utility transformer and every circuit downstream of it will be de-energized as a result:

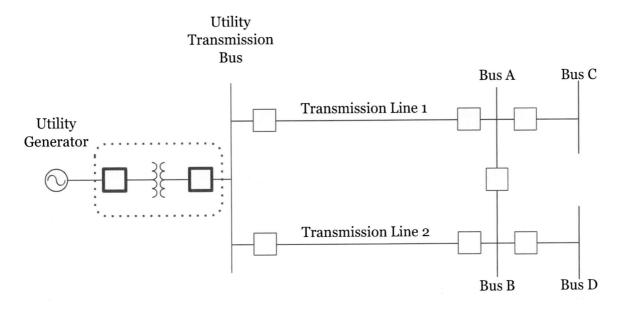

If a fault occurs within the **bus B protection zone**, the main breaker, feeder breaker, and tie breaker will operate in order to isolate the bus. Bus B and every downstream circuit of bus B will be de-energized as a result:

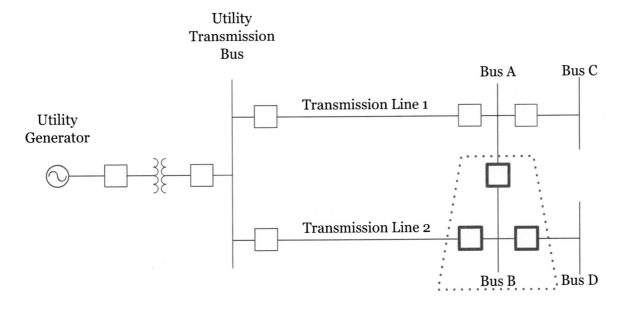

75. The answer is: (C) 1/0 AWG

The ampacity of service entrance conductors is governed by *NEC® 230.42(A)* as not less than the maximum load being served.

NEC® 310.10(H) permits conductors to be connected in parallel in sizes 1/0 AWG or greater as long as the conditions of *NEC® 310.10(H)(a), (b), and (c)* are met.

If three sets of conductors are being connected in parallel for the service entrance conductors, that means that 1/3rd of the total load will flow through each phase conductor which equals 125A per phase conductor:

375A / 3 = 125A

Since each set of phase conductors are being run in a separate raceway, we can ignore the adjustment factors required by *NEC® 310.10(H)(4)* that point to *NEC® 310.15(B)(3)(a)*.

Since there are no additional adjustment factors being applied, and not more than three current carrying conductors in each raceway, the ampacity table we will use is *NEC® Table 310.15(B)(16)*.

According to *NEC® Table 310.15(B)(16)*, 125A does not match a specific size in the 75° C column, so the next size up is 1 AWG.

However, *NEC® 310.10(H)* only permits size 1/0 AWG or greater for connecting conductors in parallel since service entrance conductors do not meet *NEC® 310.10(H) Exception No. 1*.

Because of this, we will have to round up one more time in conductor size in *NEC® Table 310.15(B)(16)* for 75° C 1/0 AWG.

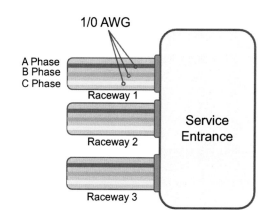

76. The answer is: (D) The protective relay becomes less sensitive as the block region increases.

The block region of an ANSI #87 differential relay is the region that the relay will not trip even if there is a mismatch of differential current present. As the block region increases, the protective relay requires a greater amount of mismatch between both differential currents in order to trip and as a result becomes less sensitive.

The opposite is also true. As the block region decreases, the protective relay requires a lower amount of mismatch between both differential currents in order to trip and as a result becomes more sensitive.

The block region can be seen when graphing the ratio of the two differential currents:

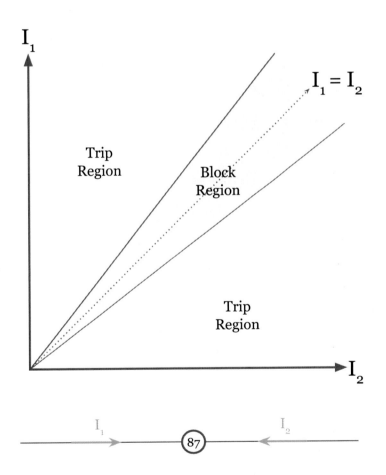

The further away from the linear line of $I_1 = I_2$ the ratio of the two currents are, the greater the mismatch that exists between them. The differential relay operates as soon as the ratio of the two currents leaves the block region and crosses into either trip region.

77. The answer is: (C) 42

First, let's draw a diagram of the three-phase transmission line:

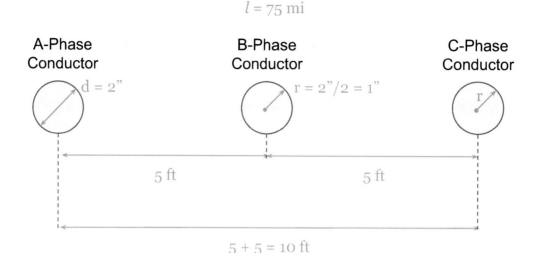

Next, let's look at the formula for the inductance ($L_{l\,3\phi}$) per foot for a three-phase transmission line.

$$L_{l\,3\phi} = 1.404(10^{-7}) \log\left(\frac{GMD}{GMR}\right)$$

In order to use this formula, we'll first need to calculate the GMD of the transmission line and the GMR of the phase conductors. Be sure to calculate both in the same units so that the units cancel when we plug these values into the above formula:

$GMD = \sqrt[3]{D_1 D_2 D_3}$
$GMD = \sqrt[3]{5ft \cdot 5ft \cdot 10ft}$
$GMD = 6.2996 ft$

$GMR = 0.7788r$
$GMR = 0.7788(1") \left(\frac{1ft}{12"}\right)$
$GMR = 0.0649 ft$

Let's use these values to solve for the inductance ($L_{l\,3\phi}$) per foot of the three-phase transmission line. Don't forget inductance (L) is in the unit of henries [H], not ohms [Ω]:

$$L_{l\,3\phi} = 1.404(10^{-7}) \log\left(\frac{6.2996 ft}{0.0649 ft}\right)$$
$$L_{l\,3\phi} = 2.7898(10^{-7}) \ H/ft$$

Next, let's multiply the inductance ($L_{l\,3\phi}$) per foot of the three-phase transmission line by the total length (l) of the transmission line to convert this value to the total inductance ($L_{3\phi}$) of the three-phase transmission line.

77. The answer is: (C) 42

(← continued from previous page)

Watch your units. Length (l) of the three-phase transmission line was given in miles, we will need to convert to feet so that the rest of the units cancel and we are left with inductance (L) in henries [H] with no length quantity:

$$L_{3\phi} = L_{l\,3\phi} \cdot l$$
$$L_{3\phi} = 2.7898(10^{-7})H/ft(75mi)\left(\frac{5,280ft}{1mi}\right)$$
$$L_{3\phi} = 0.1105H$$

The final and last step, is to convert the total inductance (L) of the three-phase transmission line in henries [H], to inductive reactance (X) in ohms [Ω]:

$$X = 2\pi f H$$
$$X = 2\pi(60Hz)(0.1105H)$$
$$X = 41.7\Omega$$

The inductive reactance (X) of the three-phase, 75 mile transmission line is 41.7 ohms. **The closest answer is 42 ohms.**

78. The answer is: (B)

Out of the four phase configurations shown, only the one shown for answer B will result in approximately 646 kVAR of reactive power for the conditions given. If each capacitor is rated for 2uF, then the reactance rating of each capacitor is 1.326kΩ:

$$X_c = \frac{1}{2\pi f C}$$

$$X_c = \frac{1}{2\pi (60Hz)(2uF)}$$

$$X_c = 1.326 k\Omega$$

Next, let's calculate the total equivalent reactance according to the answer B configuration:

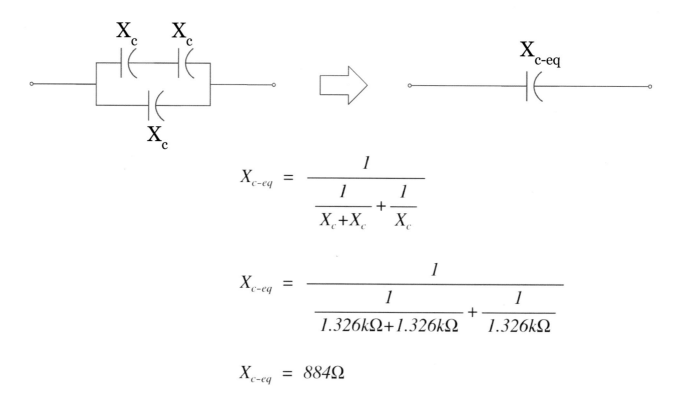

$$X_{c-eq} = \frac{1}{\frac{1}{X_c + X_c} + \frac{1}{X_c}}$$

$$X_{c-eq} = \frac{1}{\frac{1}{1.326k\Omega + 1.326k\Omega} + \frac{1}{1.326k\Omega}}$$

$$X_{c-eq} = 884\Omega$$

The total equivalent reactance of each phase of the capacitor bank is 884 ohms.

Now let's arrange each phase delta connected, and calculate the total amount of reactive power supplied by the three-phase capacitor bank, and verify that it equals the desired 646 kVAR.

78. The answer is: (B)

(← continued from previous page)

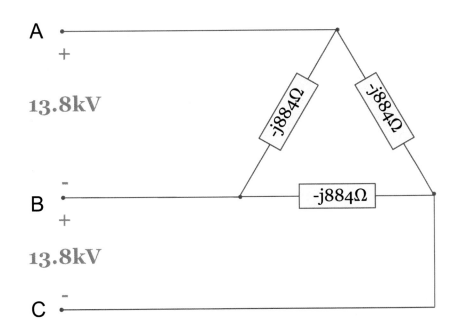

$$Q_{3\phi} = 3 \cdot Q_{1\phi}$$

$$Q_{3\phi} = 3 \cdot \frac{|V_p|^2}{X}$$

$$Q_{3\phi} = 3 \cdot \frac{13.8kV^2}{884\Omega}$$

$$Q_{3\phi} = 646 kVAR$$

The configuration shown in answer B results in the desired 646 kVAR of reactive power supplied by the three-phase capacitor bank.

79. The answer is: (C) Less reaction taking place.

As the number of cycles increases, so does corrosion and evaporation. Both of which will result in less reaction taking place.

Condensation does not build up as there is no steam created. Hydrogen is released through evaporation which lowers the electrolyte level.

Although it is possible for the cell wall to deteriorate, this not likely to happen to a well-maintained battery.

Even though performance will not decline by a significant amount until the cycle rating is reached, it will still be measurable.

80. The answer is: (B) 64

First, let's start by drawing the circuit:

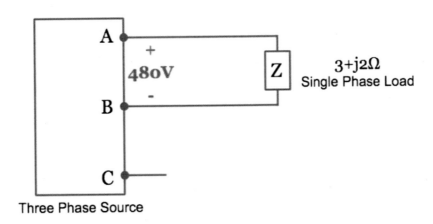

We don't know if the three-phase power source is wye or delta, but we do know that the single-phase load is connected across two terminals of the three-phase source.

This means that the voltage applied across the single-phase load is the line voltage of the three-phase source. If impedance and voltage are both known, we can calculate apparent power:

$\hat{Z} = 3+j2\Omega$

$\hat{Z} = 3.61\Omega \angle 34°$

$|S_{1\phi}| = \dfrac{|V_p|^2}{|Z|}$

$|S_{1\phi}| = \dfrac{(480V)^2}{3.61\Omega}$

$|S_{1\phi}| = 63.8kVA$

Alternatively, we can calculate apparent power using the single phase apparent power formula if we calculate the current drawn by the load first:

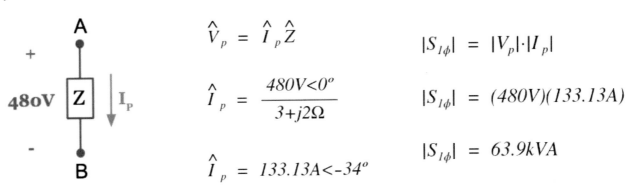

$\hat{V}_p = \hat{I}_p \hat{Z}$

$\hat{I}_p = \dfrac{480V \angle 0°}{3+j2\Omega}$

$\hat{I}_p = 133.13A \angle -34°$

$|S_{1\phi}| = |V_p| \cdot |I_p|$

$|S_{1\phi}| = (480V)(133.13A)$

$|S_{1\phi}| = 63.9kVA$

The apparent power drawn by the single phase load connected across two terminals of the three phase power source is 63.9kVA.

Answer Key

Electrical Power PE Practice Exam
10. Answer Key

Morning Session

#	Topic	Answer
1	Ch. 4 Analysis	B
2	Ch. 3 Codes and Standards	D
3	Ch. 5 Devices and Power Electronic Circuits	B
4	Ch. 3 Codes and Standards	C
5	Ch. 9 Protection	D
6	Ch. 9 Protection	A
7	Ch. 3 Codes and Standards	C
8	Ch. 4 Analysis	D
9	Ch. 6 Induction and Synchronous Machines	B
10	Ch. 6 Induction and Synchronous Machines	B
11	Ch. 9 Protection	D
12	Ch. 4 Analysis	D
13	Ch. 8 Power System Analysis	C
14	Ch. 9 Protection	C
15	Ch. 7 Electric Power Devices	A
16	Ch. 8 Power System Analysis	B
17	Ch. 3 Codes and Standards	B
18	Ch. 2 Applications	D
19	Ch. 2 Applications	A
20	Ch. 8 Power System Analysis & NEC®	B
21	Ch. 4 Analysis	A
22	Ch. 1 Measurement and Instrumentation	C
23	Ch. 9 Protection	B
24	Ch. 6 Induction and Synchronous Machines	A
25	Ch. 5 Devices and Power Electronic Circuits	D
26	Ch. 9 Protection	B
27	Ch. 6 Induction and Synchronous Machines	C
28	Ch. 3 Codes and Standards	A
29	Ch. 5 Devices and Power Electronic Circuits	B
30	Ch. 2 Applications	C
31	Ch. 9 Protection	C
32	Ch. 4 Analysis	B
33	Ch. 8 Power System Analysis	B
34	Ch. 3 Codes and Standards	D
35	Ch. 7 Electric Power Devices	A
36	Ch. 8 Power System Analysis	D
37	Ch. 7 Electric Power Devices	A
38	Ch. 8 Power System Analysis	D
39	Ch. 1 Measurement and Instrumentation	C
40	Ch. 2 Applications	A

Afternoon Session

#	Topic	Answer
41	Ch. 3 Codes and Standards	D
42	Ch. 5 Devices and Power Electronic Circuits	A
43	Ch. 7 Electric Power Devices	C
44	Ch. 7 Electric Power Devices	D
45	Ch. 5 Devices and Power Electronic Circuits	C
46	Ch. 8 Power System Analysis	B
47	Ch. 2 Applications	A
48	Ch. 9 Protection	C
49	Ch. 2 Applications	D
50	Ch. 2 Applications	C
51	Ch. 6 Induction and Synchronous Machines	C
52	Ch. 2 Applications	D
53	Ch. 3 Codes and Standards	A
54	Ch. 9 Protection	D
55	Ch. 6 Induction and Synchronous Machines	A
56	Ch. 3 Codes and Standards	B
57	Ch. 8 Power System Analysis	A
58	Ch. 7 Electric Power Devices	C
59	Ch. 8 Power System Analysis	A
60	Ch. 4 Analysis	D
61	Ch. 6 Induction and Synchronous Machines	D
62	Ch. 3 Codes and Standards	B
63	Ch. 1 Measurement and Instrumentation	D
64	Ch. 1 Measurement and Instrumentation	A
65	Ch. 3 Codes and Standards	A
66	Ch. 4 Analysis	D
67	Ch. 7 Electric Power Devices	B
68	Ch. 9 Protection	A
69	Ch. 4 Analysis	B
70	Ch. 6 Induction and Synchronous Machines	A
71	Ch. 9 Protection	A
72	Ch. 8 Power System Analysis	C
73	Ch. 5 Devices and Power Electronic Circuits	C
74	Ch. 9 Protection	B
75	Ch. 3 Codes and Standards	C
76	Ch. 9 Protection	D
77	Ch. 8 Power System Analysis	C
78	Ch. 7 Electric Power Devices	B
79	Ch. 5 Devices and Power Electronic Circuits	C
80	Ch. 4 Analysis	B

Question Order by Subject

Electrical Power PE Practice Exam
11. Question Order by Subject

How to Use the Question Order by Subject List

On the following page is the list of all questions in this practice exam organized by each of the main 9 subjects on the Electrical Power PE Exam.

Consider spending extra time reviewing material in your weaker areas as indicated by how well you scored on this practice exam, or from an official NCEES® diagnostic report from a previous attempt at the PE exam.

It is also advisable to spend extra time in the subjects that have the most number of approximate questions on the PE exam, as indicated by the NCEES® exam specifications.

For example, it would be much more beneficial to spend more time learning **protection** (Chapter 9 in the Electrical PE Review online review course) which has 13 approximate questions, followed by **codes and standards** (Chapter 3 in the Electrical PE Review online review course) which has 12 approximate questions, compared to how much time you should spend learning **measurement and instrumentation** (Chapter 3 in the Electrical PE Review online review course) which only has 4 approximate questions on the PE exam.

Electrical Power PE Practice Exam
11. Question Order by Subject

#	Subject	Ans
22	Ch. 1 Measurement and Instrumentation	C
39	Ch. 1 Measurement and Instrumentation	C
63	Ch. 1 Measurement and Instrumentation	D
64	Ch. 1 Measurement and Instrumentation	A
18	Ch. 2 Applications	D
19	Ch. 2 Applications	A
30	Ch. 2 Applications	C
40	Ch. 2 Applications	A
47	Ch. 2 Applications	A
49	Ch. 2 Applications	D
50	Ch. 2 Applications	C
52	Ch. 2 Applications	D
2	Ch. 3 Codes and Standards	D
4	Ch. 3 Codes and Standards	C
7	Ch. 3 Codes and Standards	C
17	Ch. 3 Codes and Standards	B
20	Ch. 8 Power System Analysis & NEC®	B
28	Ch. 3 Codes and Standards	A
34	Ch. 3 Codes and Standards	D
41	Ch. 3 Codes and Standards	D
53	Ch. 3 Codes and Standards	A
56	Ch. 3 Codes and Standards	B
62	Ch. 3 Codes and Standards	B
65	Ch. 3 Codes and Standards	A
75	Ch. 3 Codes and Standards	C
1	Ch. 4 Analysis	B
8	Ch. 4 Analysis	D
12	Ch. 4 Analysis	D
21	Ch. 4 Analysis	A
32	Ch. 4 Analysis	B
60	Ch. 4 Analysis	D
66	Ch. 4 Analysis	D
69	Ch. 4 Analysis	B
80	Ch. 4 Analysis	B
3	Ch. 5 Devices and Power Electronic Circuits	B
25	Ch. 5 Devices and Power Electronic Circuits	D
29	Ch. 5 Devices and Power Electronic Circuits	B
42	Ch. 5 Devices and Power Electronic Circuits	A
45	Ch. 5 Devices and Power Electronic Circuits	C
73	Ch. 5 Devices and Power Electronic Circuits	C
79	Ch. 5 Devices and Power Electronic Circuits	C
9	Ch. 6 Induction and Synchronous Machines	B
10	Ch. 6 Induction and Synchronous Machines	B
24	Ch. 6 Induction and Synchronous Machines	A
27	Ch. 6 Induction and Synchronous Machines	C
51	Ch. 6 Induction and Synchronous Machines	C
55	Ch. 6 Induction and Synchronous Machines	A
61	Ch. 6 Induction and Synchronous Machines	D
70	Ch. 6 Induction and Synchronous Machines	A
15	Ch. 7 Electric Power Devices	A
35	Ch. 7 Electric Power Devices	A
37	Ch. 7 Electric Power Devices	A
43	Ch. 7 Electric Power Devices	C
44	Ch. 7 Electric Power Devices	D
58	Ch. 7 Electric Power Devices	C
67	Ch. 7 Electric Power Devices	B
78	Ch. 7 Electric Power Devices	B
13	Ch. 8 Power System Analysis	C
16	Ch. 8 Power System Analysis	B
20	Ch. 8 Power System Analysis	B
33	Ch. 8 Power System Analysis	B
36	Ch. 8 Power System Analysis	D
38	Ch. 8 Power System Analysis	D
46	Ch. 8 Power System Analysis	B
57	Ch. 8 Power System Analysis	A
59	Ch. 8 Power System Analysis	A
72	Ch. 8 Power System Analysis	C
77	Ch. 8 Power System Analysis	C

Electrical Power PE Practice Exam
11. Question Order by Subject

5	Ch. 9 Protection	D
6	Ch. 9 Protection	A
11	Ch. 9 Protection	D
14	Ch. 9 Protection	C
23	Ch. 9 Protection	B
26	Ch. 9 Protection	B
31	Ch. 9 Protection	C
48	Ch. 9 Protection	C
54	Ch. 9 Protection	D
68	Ch. 9 Protection	A
71	Ch. 9 Protection	A
74	Ch. 9 Protection	B
76	Ch. 9 Protection	D

List of Qualitative Questions

Electrical Power PE Practice Exam
12. List of Qualitative Questions

Qualitative Questions

A **qualitative question** is one that tests on theory, definitions, practical application, and variable relationships. Little to no math is involved.

Qualitative questions continue to be an increasing concern on the PE exam as they can be quite difficult to solve compared to quantitative questions.

Quantitative questions in comparison, which make up the bulk of the PE exam, are solved by calculating a value. Even if you are not very familiar with the subject of a quantitative question, it is still possible to answer correctly as long as you are able to look up the correct formula in your references and plug in the appropriate values.

The following page contains a list of every **qualitative question** in this practice exam in order that they appear. Use this list to go directly to each **qualitative question** if you'd like to spend extra time practicing them.

For more help on the strategy of how to approach and solve **qualitative questions** on the PE exam, please refer to the following articles on www.electricalpereview.com:

1. **How to Solve Qualitative Theory Questions:**
   https://www.electricalpereview.com/thoughts-2016-october-exam-solve-qualitative-questions/

2. **Qualitative Theory Question of the Week:**
   https://www.electricalpereview.com/electrical-pe-exam-qualitative-theory-question-week/

3. **Power Supplies - Qualitative Practice Problem:**
   https://www.electricalpereview.com/qualitative-practice-problem-power-supplies/

Electrical Power PE Practice Exam
12. List of Qualitative Questions

Qualitative Questions in Order of Appearance:

#	Chapter	Ans	Type
5	Ch. 9 Protection	D	Qualitative
6	Ch. 9 Protection	A	Qualitative
10	Ch. 6 Induction and Synchronous Machines	B	Qualitative
11	Ch. 9 Protection	D	Qualitative
12	Ch. 4 Analysis	D	Qualitative
13	Ch. 8 Power System Analysis	C	Qualitative
22	Ch. 1 Measurement and Instrumentation	C	Qualitative
23	Ch. 9 Protection	B	Qualitative
24	Ch. 6 Induction and Synchronous Machines	A	Qualitative
25	Ch. 5 Devices and Power Electronic Circuits	D	Qualitative
31	Ch. 9 Protection	C	Qualitative
33	Ch. 8 Power System Analysis	B	Qualitative
36	Ch. 8 Power System Analysis	D	Qualitative
37	Ch. 7 Electric Power Devices	A	Qualitative
38	Ch. 8 Power System Analysis	D	Qualitative
45	Ch. 5 Devices and Power Electronic Circuits	C	Qualitative
52	Ch. 2 Applications	D	Qualitative
59	Ch. 8 Power System Analysis	A	Qualitative
64	Ch. 1 Measurement and Instrumentation	A	Qualitative
70	Ch. 6 Induction and Synchronous Machines	A	Qualitative
71	Ch. 9 Protection	A	Qualitative
72	Ch. 8 Power System Analysis	C	Qualitative
74	Ch. 9 Protection	B	Qualitative
76	Ch. 9 Protection	D	Qualitative
78	Ch. 7 Electric Power Devices	B	Qualitative
79	Ch. 5 Devices and Power Electronic Circuits	C	Qualitative

Made in the USA
Monee, IL
08 November 2020